T0210421

Sparse Adaptive Filters for Echo Cancellation

Synthesis Lectures on Speech and Audio Processing

Editor
B.H. Juang, *Georgia Tech*

Sparse Adaptive Filters for Echo Cancellation Constantin Paleologu, Jacob Benesty, and Silviu Ciochinˇa

ISBN: 978-3-031-01431-4   paperback
ISBN: 978-3-031-02559-4   ebook

DOI 10.1007/978-3-031-02559-4

A Publication in the Springer series
SYNTHESIS LECTURES ON SPEECH AND AUDIO PROCESSING
Lecture #6
Series Editor: B.H. Juang, *Georgia Tech*
Series ISSN
Synthesis Lectures on Speech and Audio Processing
Print 1932-121X   Electronic 1932-1678

Sparse Adaptive Filters
for Echo Cancellation

Constantin Paleologu
University Politehnica of Bucharest, Bucharest, Romania

Jacob Benesty
INRS-EMT, University of Quebec, Montreal, Canada

Silviu Ciochină
University Politehnica of Bucharest, Bucharest, Romania

SYNTHESIS LECTURES ON SPEECH AND AUDIO PROCESSING #6

ABSTRACT

Adaptive filters with a large number of coefficients are usually involved in both network and acoustic echo cancellation. Consequently, it is important to improve the convergence rate and tracking of the conventional algorithms used for these applications. This can be achieved by exploiting the sparseness character of the echo paths. Identification of sparse impulse responses was addressed mainly in the last decade with the development of the so-called "proportionate"-type algorithms. The goal of this book is to present the most important sparse adaptive filters developed for echo cancellation. Besides a comprehensive review of the basic proportionate-type algorithms, we also present some of the latest developments in the field and propose some new solutions for further performance improvement, e.g., variable step-size versions and novel proportionate-type affine projection algorithms. An experimental study is also provided in order to compare many sparse adaptive filters in different echo cancellation scenarios.

KEYWORDS

network and acoustic echo cancellation, adaptive filters, sparseness, Wiener, LMS, NLMS, VSS-NLMS, PNLMS, IPNLMS, EG±, VSS-PNLMS, APA, PAPA

Contents

CHAPTER 1

Introduction

1.1 ECHO CANCELLATION

Among the wide range of adaptive filtering applications, echo cancellation is likely the most interesting and challenging one. The original idea of this application appeared in the sixties (66), and it can be considered as a real milestone in telecommunication systems. A general scheme for echo cancellation is depicted in Fig. 1.1. In both network and acoustic echo cancellation contexts (5), the basic principle is to build a model of the echo path impulse response that needs to be identified with an adaptive filter, which provides at its output a replica of the echo (that is further subtracted from the reference signal). The main difference between these two applications is the way in which the echo arises. In the network (or electrical) echo problem, there is an unbalanced coupling between the 2-wire and 4-wire circuits which results in echo, while the acoustic echo is due to the acoustic coupling between the microphone and loudspeaker (e.g., as in speakerphones). However, in both cases, the adaptive filter has to model an unknown system, i.e., the echo path.

It is interesting to notice that the scheme from Fig. 1.1 can be interpreted as a combination of two classes of adaptive system configurations, according to the adaptive filter theory (33). First, it represents a "system identification" configuration because the goal is to identify an unknown system (i.e., the echo path) with its output corrupted by an apparently "undesired" signal (i.e., the near-end signal). But it also can be viewed as an "interference cancelling" configuration, aiming to recover a "useful" signal (i.e., the near-end signal) corrupted by an undesired perturbation (i.e., the echo signal); consequently, the "useful" signal should be recovered from the error signal of the adaptive filter.

Even if the formulation of the echo cancellation problem is straightforward, its specific features represent a challenge for any adaptive algorithm. There are several issues associated with this application, and they are as follows. First, the echo paths can have excessive lengths in time, e.g., up to hundreds of milliseconds. In network echo cancellation, the usual lengths are in the range between 32 and 128 milliseconds, while in acoustic echo cancellation, these lengths can be even higher. Consequently, long length adaptive filters are required (hundreds or even thousands of coefficients), influencing the convergence rate of the algorithm. Besides, the echo paths are time-variant systems, requiring good tracking capabilities for the echo canceller. Second, the echo signal is combined with the near-end signal; ideally, the adaptive filter should separate this mixture and provide an estimate of the echo at its output as well as an estimate of the near-end from the error signal. This is not an easy task since the near-end signal can contain both the background noise and the near-end speech; the background noise can be non-stationary and strong while the near-end speech acts like

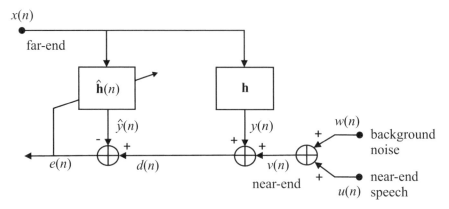

Figure 1.1: General configuration for echo cancellation.

a large level disturbance. Last but not least, the input of the adaptive filter (i.e., the far-end signal) is mainly speech, which is a non-stationary and highly correlated signal that can influence the overall performance of adaptive algorithms.

Each of the previously addressed problems implies some special requirements for the adaptive algorithms used for echo cancellation. Summarizing, the "ideal" algorithm should have a high convergence rate and good tracking capabilities (in order to deal with the high length and time-varying nature of the echo path impulse responses) but achieving low misadjustment. These issues should be obtained despite the non-stationary character of the input signal (i.e., speech). Also, the algorithm should be robust against the near-end signal variations, e.g., background noise variations and double talk. Finally, its computational complexity should be moderate, providing both efficient and low-cost real-time implementation. Even if the literature of adaptive filters contains a lot of very interesting and useful algorithms (33), there is not an adaptive algorithm that satisfies all the previous requirements.

Different types of adaptive filters have been involved in the context of echo cancellation. One of the most popular is the normalized least-mean-square (NLMS) algorithm. The main reasons behind this popularity are its moderate computational complexity, together with its good numerical stability. Also, the affine projection algorithm (APA) [originally proposed in (54)] and some of its versions, e.g., (24), (70), were found to be very attractive choices for echo cancellation. However, there is still a need to improve the performance of these algorithms for echo cancellation. More importantly, it is necessary to find some way to increase the convergence rate and tracking of the algorithms since it is known that the performance of both NLMS and APA are limited for high length adaptive filters. This can be partially overcome by exploiting the character of the system to be identified (i.e., the echo path) as it will be explained in Section 1.3.

1.2   DOUBLE-TALK DETECTION

One of the most challenging problems in echo cancellation is the double-talk situation, i.e., the talkers on both sides speak simultaneously. The behavior of the adaptive filter can be seriously affected in this case, up to divergence. For this reason, the echo canceller is usually equipped with a double-talk detector (DTD), in order to slow down or completely halt the adaptation process during double-talk periods (5). The main challenge for the DTD algorithm is to "feel" the presence of the near-end speech.

A lot of very interesting DTD algorithms have been proposed. Maybe the simplest one is the well-known Geigel DTD (17), which provides an efficient and low-complexity solution, especially for network echo cancellation. Other more complex algorithms, e.g., based on coherence and cross-correlation methods (4), (22), have proven to give better results but with an increased computational complexity. Nevertheless, there is some inherent delay in the decision of any DTD; during this small period, a few undetected large amplitude samples can perturb the echo path estimate considerably. Consequently, it is highly desirable to improve the robustness of adaptive algorithms in order to handle a certain amount of double talk without diverging. This is the motivation behind the development of the so-called robust algorithms. A solution of this kind, based on the theory of robust statistics (39), was proposed in (23). More recent frameworks for designing robust adaptive algorithms can be found in (62) and (63). Other approaches try to minimize or even annihilate the role of the DTD, e.g., using a postfilter to suppress the residual echo that remains after cancellation (19), or using an adaptive cross-spectral technique instead of an adaptive algorithm (46), (53).

It is known that the adaptive filter can "help" in double-talk situations by using a very small (i.e., close to zero) step-size parameter, which prevents the divergence. However, in this case, the convergence rate and the tracking capabilities of the algorithm will suffer a lot; it is also known that a high value of the step size is needed to accomplish these performance criteria. Consequently, the step-size parameter should be chosen based on a compromise between fast convergence rate and good tracking capabilities on the one hand, and low misadjustment and double-talk robustness on the other hand. In order to meet this conflicting requirement, a practical solution is to use a variable step-size (VSS) approach (47). A number of VSS-NLMS algorithms and VSS-APAs were developed, e.g., (9), (55), (56), (61), (65). Nevertheless, some of these algorithms require the tuning of some parameters which are difficult to control in practice. For real-world echo cancellation, it is highly desirable to use non-parametric algorithms, in the sense that no information about the environment is required, e.g., (55), (56). These algorithms are developed based on another objective of echo cancellation, i.e., to recover the near-end signal from the error signal of the adaptive filter. As a consequence, these VSS algorithms are equipped with good robustness features against near-end signal variations, like double talk.

1.3   SPARSE ADAPTIVE FILTERS

As we have mentioned in Section 1.1, the main goal in echo cancellation is to identify an unknown system, i.e., the echo path, providing at the output of the adaptive filter a replica of the echo. Consequently, this application is basically a "system identification" problem (33). Nevertheless, the echo paths (for both network and acoustic echo cancellation scenarios) have a specific property, which can be used in order to help the adaptation process. These systems are sparse in nature, i.e., a small percentage of the impulse response components have a significant magnitude while the rest are zero or small. In the case of the network echo, the bulk delay can vary in the range between 32 and 128 milliseconds (depending on the network conditions), while the "active" region is in the range between 8 and 12 milliseconds; consequently, the impulse response of the network echo path contains mainly "inactive" coefficients (i.e., close to zero). The sparseness of an acoustic impulse response is more problematic because it depends on many factors, e.g., reverberation time, the distance between loudspeaker and microphone, different changes in the environment (e.g., temperature or pressure); however, the acoustic echo paths are in general less sparse as compared to their network counterparts, but their sparseness can also be exploited.

The sparseness character of the echo paths inspired the idea to "proportionate" the algorithm behavior, i.e., to update each coefficient of the filter independently of the others, by adjusting the adaptation step size in proportion to the magnitude of the estimated filter coefficient. In this manner, the adaptation gain is "proportionately" redistributed among all the coefficients, emphasizing the large ones in order to speed up their convergence and, consequently, to increase the overall convergence rate. Even if the idea of exploiting the sparseness character of the systems has appeared in the nineties, e.g., (35), (48), (69), the proportionate NLMS (PNLMS) algorithm (18) proposed, by Duttweiler a decade ago, was one of the first "true" proportionate-type algorithms and maybe the most referred one. As compared to its predecessors, the update rule of the PNLMS algorithm is based only on the current adaptive filter estimate, requiring no a priori information about the echo path. However, the PNLMS algorithm was developed in an "intuitively" manner, because the equations used to calculate the step-size control factors are not based on any optimization criteria but are designed in an ad-hoc way. For this reason, after an initial fast convergence phase, the convergence rate of the PNLMS algorithm significantly slows down. Besides, it is sensitive to the sparseness degree of the system, i.e., the convergence rate is reduced when the echo paths are not very sparse.

In order to deal with these problems, many proportionate-type algorithms were developed in the last decade. The overall goal of this book is to present and analyze the most important sparse adaptive filters, in order to outline their capabilities and performances in the context of echo cancellation. To facilitate the flow of the book, the basic notions regarding the sparseness and performance measures are presented in Chapters 2 and 3, respectively. Also, Chapter 4 reviews the Wiener and basic adaptive filters, including the steepest-descend method and the stochastic algorithms. After these introductory chapters, the basic proportionate-type NLMS adaptive filters are presented in Chapter 5; the classical PNLMS (18), the improved PNLMS (6), and other related algorithms are discussed in this chapter. The exponentiated gradient (EG) algorithms (41) and their connections with the basic sparse adaptive filters are presented in Chapter 6. Some of the most recent developments in the field, including the mu-law (14), (15) and other new PNLMS-type algorithms are included in Chapter 7. A variable step-size PNLMS-type algorithm is developed in Chapter 8, aiming to better compromise between the conflicting requirements of fast convergence and low misadjustment encountered in the classical versions. Chapter 9 is dedicated to the family of proportionate-type APAs (PAPAs), which further improve the performance of the PNLMS-type algorithms. Finally, an experimental study is presented in Chapter 10, comparing these algorithms in different echo cancellation scenarios.

1.4 NOTATION

Throughout this text, the following notation will be used (see also Fig. 1.1):

- n is the discrete-time index,

- superscript ^{T} denotes transpose of a vector or a matrix,

- L is the length of the adaptive filter (assumed to be equal to the length of the echo path),

- $x(n)$ is the far-end signal (i.e., the input signal of the adaptive filter and loudspeaker),

- $\mathbf{x}(n) = \begin{bmatrix} x(n) & x(n-1) & \cdots & x(n-L+1) \end{bmatrix}^T$ is a vector containing the most recent L samples of the input signal,

- $\mathbf{h} = \begin{bmatrix} h_0 & h_1 & \cdots & h_{L-1} \end{bmatrix}^T$ is the impulse response of the system (i.e., the echo path),

- $\hat{\mathbf{h}}(n) = \begin{bmatrix} \hat{h}_0(n) & \hat{h}_1(n) & \cdots & \hat{h}_{L-1}(n) \end{bmatrix}^T$ is the estimated impulse response at time n (i.e., the adaptive filter at time n),

- $y(n) = \mathbf{h}^T \mathbf{x}(n)$ is the echo signal,

- $\hat{y}(n) = \hat{\mathbf{h}}^T(n-1)\mathbf{x}(n)$ is the output of the adaptive filter at time n (i.e., the estimated echo),

- $w(n)$ is the background noise at the near-end,

- $u(n)$ is the near-end speech (in case of double talk),

- $v(n) = w(n) + u(n)$ is the near-end signal,

- $d(n) = y(n) + v(n)$ is the reference signal (also known as the desired signal), but most of the time we will not consider the near-end speech so that $d(n) = y(n) + w(n)$,

- $e(n) = d(n) - \hat{\mathbf{h}}^T(n-1)\mathbf{x}(n)$ is the a priori error signal,

- $\varepsilon(n) = d(n) - \hat{\mathbf{h}}^T(n)\mathbf{x}(n)$ is the a posteriori error signal,

- μ is the step-size parameter of the adaptive algorithm,

- α is the normalized step-size parameter of the adaptive algorithm,

- δ is the regularization constant,

- $\mathbf{G}(n)$ is a diagonal matrix $(L \times L)$ containing the "proportionate" factors at time n,

- $\mathbf{g}(n) = \begin{bmatrix} g_0(n) & g_1(n) & \cdots & g_{L-1}(n) \end{bmatrix}^T$ is a vector containing the diagonal elements of $\mathbf{G}(n)$,

- P is the projection order of the affine projection algorithm,

- $\mathbf{X}(n) = \begin{bmatrix} \mathbf{x}(n) & \mathbf{x}(n-1) & \cdots & \mathbf{x}(n-P+1) \end{bmatrix}$ is the input signal matrix $(L \times P)$,

- $\mathbf{d}(n) = \begin{bmatrix} d(n) & d(n-1) & \cdots & d(n-P+1) \end{bmatrix}^T$ is a vector containing the most recent P samples of the reference signal,

- $\mathbf{e}(n) = \mathbf{d}(n) - \mathbf{X}^T(n)\hat{\mathbf{h}}(n-1)$ is the a priori error signal vector at time n,

- $\boldsymbol{\varepsilon}(n) = \mathbf{d}(n) - \mathbf{X}^T(n)\hat{\mathbf{h}}(n)$ is the a posteriori error signal vector at time n.

CHAPTER 2

Sparseness Measures

Impulse responses may be very much different from one to another in networks or rooms; their characteristic depends on many factors, so it is important to be able to quantify how sparse or dense they are. In this chapter, we discuss some sparseness measures that can reliably quantify the sparseness of a vector.

We believe that a good sparseness measure needs to have the following properties (38):

- bounded rather than infinite range of definition,

- invariant with a non-zero scaling factor, and

- independent of the sorting order of the channel impulse response coefficients.

The first two properties are easy to understand. The third implies that if we sort the channel impulse response coefficients in different manners, the sparseness measure will not be any different. This makes sense, and it is important since sparseness is all about the dynamic range of the impulse response coefficients and has nothing to do with their order.

2.1 VECTOR NORMS

Many vector norms (29) exist in the literature, but four of them are of great interest to us.

Consider the vector

$$\mathbf{h} = \begin{bmatrix} h_0 & h_1 & \cdots & h_{L-1} \end{bmatrix}^T \neq \mathbf{0} \tag{2.1}$$

of length $L > 1$ and define the function

$$f(h_l) = \begin{cases} 1, & h_l \neq 0 \\ 0, & h_l = 0 \end{cases}, \tag{2.2}$$

then the ℓ_0 norm of $\mathbf{h}$ is (16)

$$\|\mathbf{h}\|_0 = \sum_{l=0}^{L-1} f(h_l). \tag{2.3}$$

Basically, the ℓ_0 norm of a vector is equal to the number of its non-zero components. For $\mathbf{h} \neq \mathbf{0}$, we always have

$$1 \leq \|\mathbf{h}\|_0 \leq L. \tag{2.4}$$

The fact that $f(h_l)$ is not a continuous function and that many elements of the vector $\mathbf{h}$ can be very small but not exactly zero, makes the ℓ_0 norm as defined in (2.3) difficult to use in practice and is often approximated by continuous functions.

The ℓ_1, ℓ_2, and ℓ_∞ (maximum) norms (29) of the vector $\mathbf{h}$ are defined as, respectively,

$$\|\mathbf{h}\|_1 = \sum_{l=0}^{L-1} |h_l|, \tag{2.5}$$

$$\|\mathbf{h}\|_2 = \sqrt{\sum_{l=0}^{L-1} h_l^2}$$
$$= \sqrt{\mathbf{h}^T \mathbf{h}}, \tag{2.6}$$

and

$$\|\mathbf{h}\|_\infty = \max_{0 \le l \le L-1} |h_l|. \tag{2.7}$$

It can be shown that (29)

$$1 \le \frac{\|\mathbf{h}\|_1}{\|\mathbf{h}\|_2} \le \sqrt{L}, \tag{2.8}$$

$$1 \le \frac{\|\mathbf{h}\|_1}{\|\mathbf{h}\|_\infty} \le L, \tag{2.9}$$

$$1 \le \frac{\|\mathbf{h}\|_2}{\|\mathbf{h}\|_\infty} \le \sqrt{L}. \tag{2.10}$$

These inequalities are very important for the derivation of sparseness measures since the ratios of different vector norms are lower and upper bounded by values independent of the characteristic of the vector.

Since

$$\frac{|h_l|}{\sqrt{\sum_{l=0}^{L-1} h_l^2}} \le f(h_l), \ \forall l, \tag{2.11}$$

$$\frac{|h_l|}{\|\mathbf{h}\|_\infty} \le f(h_l), \ \forall l, \tag{2.12}$$

and

$$\frac{h_l^2}{\left(\sum_{l=0}^{L-1} |h_l|\right)^2} \le f(h_l), \ \forall l, \tag{2.13}$$

we deduce that

$$\frac{1}{\sqrt{\|\mathbf{h}\|_0}} \leq \frac{\|\mathbf{h}\|_1}{\|\mathbf{h}\|_2} \leq \|\mathbf{h}\|_0 , \tag{2.14}$$

$$\frac{1}{\sqrt{\|\mathbf{h}\|_0}} \leq \frac{\|\mathbf{h}\|_1}{\|\mathbf{h}\|_\infty} \leq \|\mathbf{h}\|_0 . \tag{2.15}$$

2.2   EXAMPLES OF IMPULSE RESPONSES

Before deriving different sparseness measures, we present three symbolic filters:

- the Dirac filter,

$$\mathbf{h}_{\mathrm{d}} = \begin{bmatrix} 1 & 0 & \cdots & 0 \end{bmatrix}^T , \tag{2.16}$$

- the uniform filter,

$$\mathbf{h}_{\mathrm{u}} = \begin{bmatrix} 1 & 1 & \cdots & 1 \end{bmatrix}^T , \tag{2.17}$$

- and the exponentially decaying filter,

$$\mathbf{h}_{\mathrm{e}} = \begin{bmatrix} 1 & \exp\left(-\frac{1}{\beta}\right) & \cdots & \exp\left(-\frac{L-1}{\beta}\right) \end{bmatrix}^T , \tag{2.18}$$

  where β is a positive number called the decay constant.

The Dirac and uniform filters are actually two particular cases of the exponentially decaying filter:

$$\lim_{\beta \to 0} \mathbf{h}_{\mathrm{e}} = \mathbf{h}_{\mathrm{d}}, \tag{2.19}$$

$$\lim_{\beta \to \infty} \mathbf{h}_{\mathrm{e}} = \mathbf{h}_{\mathrm{u}}. \tag{2.20}$$

While the Dirac filter is the sparsest of all possible impulse responses, the uniform filter is the densest or least sparse one. The filter $\mathbf{h}_{\mathrm{e}}$ is a good model of acoustic impulse responses where β depends on the reverberation time. For a long reverberation time (large β), $\mathbf{h}_{\mathrm{e}}$ will decay slowly while for a short reverberation time (small β), $\mathbf{h}_{\mathrm{e}}$ will decay rapidly.

2.3   SPARSENESS MEASURE BASED ON THE ℓ_0 NORM

The most obvious sparseness measure is based on the ℓ_0 norm and is defined as

$$\xi_0(\mathbf{h}) = \frac{L}{L-1} \left(1 - \frac{\|\mathbf{h}\|_0}{L} \right). \tag{2.21}$$

We see that the closer the measure is to 1, the sparser is the impulse response. On the contrary, the closer the measure is to 0, the denser or less sparse is the impulse response.

For the sparseness measure given in (2.21), we have the following properties:

$$
\begin{array}{lll}
\text{(a)} & 0 \leq \xi_0(\mathbf{h}) \leq 1, & (2.22) \\
\text{(b)} & \forall a \neq 0, \ \ \xi_0(a\mathbf{h}) = \xi_0(\mathbf{h}), & (2.23) \\
\text{(c)} & \xi_0(\mathbf{h}_d) = 1, & (2.24) \\
\text{(d)} & \xi_0(\mathbf{h}_u) = 0. & (2.25)
\end{array}
$$

We see from these properties that the measure is bounded and is not affected by a non-zero scaling factor. Furthermore, since the ℓ_0 norm of a vector is independent of the order of the vector coefficients, so is the defined sparseness measure.

While $\xi_0(\mathbf{h})$ is interesting from a theoretical point of view, it may be very limited in practice since the coefficients of acoustic and network impulse responses are rarely exactly equal to zero even though many of them can be very small. In particular, the value of $\xi_0(\mathbf{h}_e)$ does not make that much sense in our context. Therefore, the ℓ_0 norm to measuring sparseness is the best possible choice only when the coefficients of the impulse response are 0's and 1's; however, this simple scenario does not occur, in general, in the applications of echo cancellation.

2.4 SPARSENESS MEASURE BASED ON THE ℓ_1 AND ℓ_2 NORMS

A more appropriate sparseness measure should be:

- equal to 0 if all components of the vector are equal (up to a sign factor),

- equal to 1 if the vector contains only one non-zero component,

- able to interpolate smoothly between the two extremes.

A measure depending on the ℓ_1 and ℓ_2 norms already exists. It is defined as (37), (38)

$$
\xi_{12}(\mathbf{h}) = \frac{L}{L - \sqrt{L}} \left(1 - \frac{\|\mathbf{h}\|_1}{\sqrt{L}\|\mathbf{h}\|_2} \right). \tag{2.26}
$$

We have the following properties:

$$
\begin{array}{lll}
\text{(a)} & 0 \leq \xi_{12}(\mathbf{h}) \leq 1, & (2.27) \\
\text{(b)} & \forall a \neq 0, \ \ \xi_{12}(a\mathbf{h}) = \xi_{12}(\mathbf{h}), & (2.28) \\
\text{(c)} & \xi_{12}(\mathbf{h}_d) = 1, & (2.29) \\
\text{(d)} & \xi_{12}(\mathbf{h}_u) = 0. & (2.30)
\end{array}
$$

These properties are identical to the ones for $\xi_0(\mathbf{h})$. The fundamental difference is that now $\xi_{12}(\mathbf{h}_e)$ varies smoothly between 1 and 0, depending on the reverberation time (or β) as shown in Fig. 2.1 where the length of $\mathbf{h}_e$ is $L = 256$ and the decay constant (β) varies from 1 to 100.

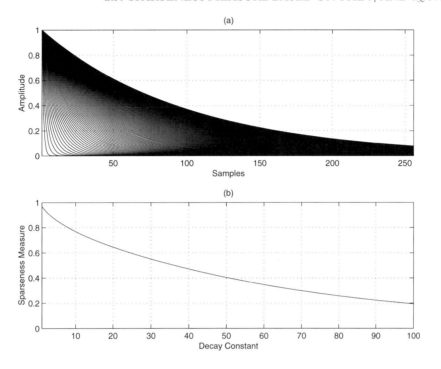

Figure 2.1: Values of the sparseness measure, ξ_{12}, for the exponentially decaying filter, $\mathbf{h}_\mathrm{e}$, with various decay constants. (a) Impulse responses $\mathbf{h}_\mathrm{e}$ of length $L = 256$ for values of the decay constant β from 1 to 100. (b) Sparseness measure for $\mathbf{h}_\mathrm{e}$ as a function of the decay constant, β.

2.5 SPARSENESS MEASURE BASED ON THE ℓ_1 AND ℓ_∞ NORMS

We define the sparseness measure based on the ℓ_1 and ℓ_∞ norms as

$$\xi_{1\infty}(\mathbf{h}) = \frac{L}{L-1}\left(1 - \frac{\|\mathbf{h}\|_1}{L\|\mathbf{h}\|_\infty}\right), \tag{2.31}$$

which also has the desirable properties:

(a)	$0 \leq \xi_{1\infty}(\mathbf{h}) \leq 1,$	(2.32)
(b)	$\forall a \neq 0, \;\; \xi_{1\infty}(a\mathbf{h}) = \xi_{1\infty}(\mathbf{h}),$	(2.33)
(c)	$\xi_{1\infty}(\mathbf{h}_\mathrm{d}) = 1,$	(2.34)
(d)	$\xi_{1\infty}(\mathbf{h}_\mathrm{u}) = 0.$	(2.35)

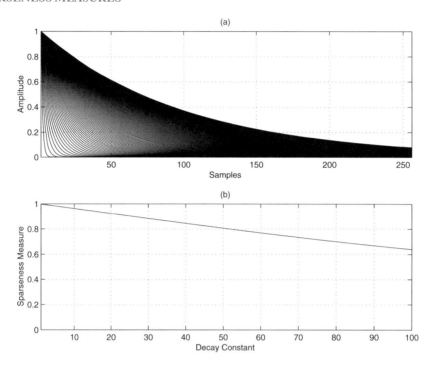

Figure 2.2: Values of the sparseness measure, $\xi_{1\infty}$, for the exponentially decaying filter, $\mathbf{h}_e$, with various decay constants. Other conditions the same as in Fig. 2.1.

It is easy to verify that

$$\xi_{1\infty}(\mathbf{h}_e) = \frac{L}{L-1}\left\{1 - \frac{1 - \exp\left(-\frac{L}{\beta}\right)}{L\left[1 - \exp\left(-\frac{1}{\beta}\right)\right]}\right\}. \tag{2.36}$$

Here again, $\xi_{1\infty}(\mathbf{h}_e)$ varies smoothly between 1 and 0 as shown in Fig. 2.2 but less rapidly than $\xi_{12}(\mathbf{h}_e)$.

2.6   SPARSENESS MEASURE BASED ON THE ℓ_2 AND ℓ_∞ NORMS

A sparseness measure depending on the ℓ_2 and ℓ_∞ norms can also be found following the definitions of $\xi_{12}(\mathbf{h})$ and $\xi_{1\infty}(\mathbf{h})$; we define it as

$$\xi_{2\infty}(\mathbf{h}) = \frac{L}{L-\sqrt{L}}\left(1 - \frac{\|\mathbf{h}\|_2}{\sqrt{L}\|\mathbf{h}\|_\infty}\right) \tag{2.37}$$

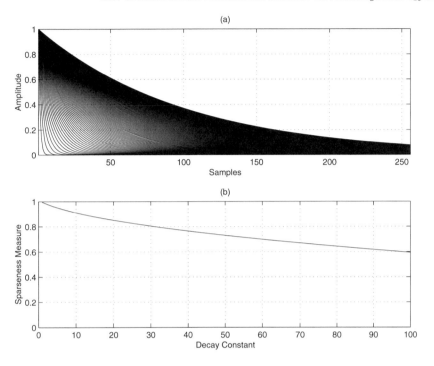

Figure 2.3: Values of the sparseness measure, $\xi_{2\infty}$, for the exponentially decaying filter, $\mathbf{h}_e$, with various decay constants. Other conditions the same as in Fig. 2.1.

and has the following properties:

$$\text{(a)} \qquad 0 \le \xi_{2\infty}(\mathbf{h}) \le 1, \tag{2.38}$$

$$\text{(b)} \qquad \forall a \ne 0, \;\; \xi_{2\infty}(a\mathbf{h}) = \xi_{2\infty}(\mathbf{h}), \tag{2.39}$$

$$\text{(c)} \qquad \xi_{2\infty}(\mathbf{h}_d) = 1, \tag{2.40}$$

$$\text{(d)} \qquad \xi_{2\infty}(\mathbf{h}_u) = 0. \tag{2.41}$$

From Fig. 2.3 we see that the behavior of $\xi_{2\infty}(\mathbf{h}_e)$ is similar to the behavior of $\xi_{1\infty}(\mathbf{h}_e)$. Intuitively, we believe that the values of $\xi_{12}(\mathbf{h}_e)$ are good representations of sparse impulse responses while values of $\xi_{1\infty}(\mathbf{h}_e)$ and $\xi_{2\infty}(\mathbf{h}_e)$ are good representations of dense impulse responses. We can combine differently these three sparseness measures to try having a better grip of a particular scenario. For example,

$$\xi_{12\infty}(\mathbf{h}) = \frac{\xi_{12}(\mathbf{h}) + \xi_{2\infty}(\mathbf{h})}{2} \tag{2.42}$$

can be another good measure of sparseness.

Finally, to conclude this chapter, it is easy to check that the three sparseness measures $\xi_{12}(\mathbf{h})$, $\xi_{1\infty}(\mathbf{h})$, and $\xi_{2\infty}(\mathbf{h})$ are related as follows:

$$\left[1 - \left(1 - \frac{1}{\sqrt{L}}\right)\xi_{12}(\mathbf{h})\right]\left[1 - \left(1 - \frac{1}{\sqrt{L}}\right)\xi_{2\infty}(\mathbf{h})\right] = 1 - \left(1 - \frac{1}{L}\right)\xi_{1\infty}(\mathbf{h}). \qquad (2.43)$$

CHAPTER 3

Performance Measures

In the echo cancellation problem, everything is about how much the undesired echo is attenuated. There are many (implicit or explicit) ways to measure this attenuation, but three performance measures are very common in the contexts of adaptive filtering, identification, and signal cancellation. They are, by far, the most used ones in the literature and are explained in this chapter.

3.1 MEAN-SQUARE ERROR

The mean-square error (MSE) is the mean-square value of the difference between the desired signal and the filter output (33). It is defined as

$$\text{MSE}(n) = E\left\{\left[d(n) - \hat{y}(n)\right]^2\right\}, \tag{3.1}$$

where $E\{\cdot\}$ denotes mathematical expectation,

$$\begin{aligned} d(n) &= y(n) + w(n) \\ &= \mathbf{h}^T\mathbf{x}(n) + w(n) \end{aligned} \tag{3.2}$$

is the desired signal (i.e., the echo plus noise), and

$$\hat{y}(n) = \hat{\mathbf{h}}^T(n-1)\mathbf{x}(n) \tag{3.3}$$

is the output of the adaptive filter at time n.

Developing (3.1) and assuming that $\hat{\mathbf{h}}(n)$ is deterministic, we obtain

$$\text{MSE}(n) = \left[\mathbf{h} - \hat{\mathbf{h}}(n-1)\right]^T \mathbf{R}_\mathbf{x}\left[\mathbf{h} - \hat{\mathbf{h}}(n-1)\right] + \sigma_w^2, \tag{3.4}$$

where

$$\mathbf{R}_\mathbf{x} = E\left[\mathbf{x}(n)\mathbf{x}^T(n)\right] \tag{3.5}$$

is the correlation matrix of $\mathbf{x}(n)$ and

$$\sigma_w^2 = E\left[w^2(n)\right] \tag{3.6}$$

is the variance of the noise. From (3.4), we observe that when the adaptive filter, $\hat{\mathbf{h}}(n)$, converges to the impulse response, $\mathbf{h}$, of the system, the MSE converges to the variance of the noise, σ_w^2, i.e.,

$$\lim_{n\to\infty} \text{MSE}(n) = \sigma_w^2. \tag{3.7}$$

The MSE may not be the best performance measure for two reasons. First, in some situations or applications even if the MSE converges to the level of the noise, or to small values, does not mean that $\hat{\mathbf{h}}(n)$ converges to $\mathbf{h}$. A typical example of this behavior is the problem of stereophonic acoustic echo cancellation [see (3), (5), and (67) for more details]. Second, the MSE does not give an explicit or precise measure of the echo attenuation since it also depends on the variance of the additive noise.

3.2   ECHO-RETURN LOSS ENHANCEMENT

A more objective measure to assess the echo cancellation by the adaptive filter is the echo-return loss enhancement (ERLE), defined as (32)

$$
\begin{aligned}
\text{ERLE}(n) &= \frac{E\left[y^2(n)\right]}{E\left\{\left[y(n) - \hat{y}(n)\right]^2\right\}} \\
&= \frac{\mathbf{h}^T \mathbf{R_x} \mathbf{h}}{\left[\mathbf{h} - \hat{\mathbf{h}}(n-1)\right]^T \mathbf{R_x}\left[\mathbf{h} - \hat{\mathbf{h}}(n-1)\right]}.
\end{aligned}
\tag{3.8}
$$

In the second line of the previous expression, we assumed that $\hat{\mathbf{h}}(n)$ is deterministic. We observe from (3.8) that the ERLE does not depend on the additive noise. Furthermore, when $\hat{\mathbf{h}}(n)$ converges to $\mathbf{h}$, the ERLE goes to infinity. Therefore, the larger is the ERLE, the more the echo is attenuated.

It is straightforward to see that the MSE and ERLE are related as follows

$$
\text{MSE}(n) = \sigma_w^2 + \frac{\mathbf{h}^T \mathbf{R_x} \mathbf{h}}{\text{ERLE}(n)}.
\tag{3.9}
$$

Another interesting way to write the ERLE is

$$
\text{ERLE}(n) = \frac{\text{ENR}}{\text{RENR}(n-1)},
\tag{3.10}
$$

where

$$
\text{ENR} = \frac{\mathbf{h}^T \mathbf{R_x} \mathbf{h}}{\sigma_w^2}
\tag{3.11}
$$

is the echo-to-noise ratio (ENR) and

$$
\text{RENR}(n-1) = \frac{\left[\mathbf{h} - \hat{\mathbf{h}}(n-1)\right]^T \mathbf{R_x}\left[\mathbf{h} - \hat{\mathbf{h}}(n-1)\right]}{\sigma_w^2}
\tag{3.12}
$$

is the residual-echo-to-noise ratio (RENR). Low ENRs usually affect the performance of adaptive algorithms and, as a result, the RENR will also be affected.

3.3   MISALIGNMENT

Probably the most used performance measure in echo cancellation is the so-called misalignment (5). It quantifies directly how "well" (in terms of convergence, tracking, and accuracy to the solution) an adaptive filter converges to the impulse response of the system that needs to be identified. The misalignment is defined as

$$\text{Mis}(n) = \frac{\left\| \mathbf{h} - \hat{\mathbf{h}}(n) \right\|_2^2}{\left\| \mathbf{h} \right\|_2^2}, \tag{3.13}$$

or in dB,

$$\text{Mis}(n) = 20 \log_{10} \frac{\left\| \mathbf{h} - \hat{\mathbf{h}}(n) \right\|_2}{\left\| \mathbf{h} \right\|_2} \quad \text{(dB)}. \tag{3.14}$$

If the far-end signal, $x(n)$, is white, then $\mathbf{R_x} = \sigma_x^2 \mathbf{I}$, where

$$\sigma_x^2 = E\left[x^2(n) \right] \tag{3.15}$$

is the variance of $x(n)$ and $\mathbf{I}$ is the identity matrix of size $L \times L$. As a result, we deduce a very simple relationship between the ERLE and the misalignment, which is

$$\text{ERLE}(n) = \frac{1}{\text{Mis}(n-1)}. \tag{3.16}$$

C H A P T E R   4

Wiener and Basic Adaptive Filters

The Wiener filter has been an extremely useful tool since its invention in the early 30's by Norbert Wiener (76) and is very popular in adaptive filtering, in general, and in echo cancellation, in particular. The objective of this chapter is to present the most fundamental results of the Wiener theory with an emphasis on the Wiener-Hopf equations that can lead to an optimal estimation of the impulse response of the system, but these equations are not convenient to solve in practice. An alternative approach to solving these equations is via an adaptive filter, which relies on new data at each time iteration for an estimation of the optimal solution. That is why this part also describes the most classical adaptive algorithms that are able to converge, in a reasonable amount of time, to the optimal Wiener filter.

4.1   WIENER FILTER

With the Wiener theory, it is possible to identify the impulse response $\mathbf{h}$, given $x(n)$ and $d(n)$. Define the error signal

$$
\begin{aligned}
e(n) &= d(n) - \hat{y}(n) \\
&= d(n) - \hat{\mathbf{h}}^T \mathbf{x}(n),
\end{aligned}
\tag{4.1}
$$

where $\hat{\mathbf{h}}$ is an estimate of $\mathbf{h}$ (and both vectors have the same length L).

To find the optimal filter, we need to minimize a cost function which is always built around the error signal [eq. (4.1)]. The usual choice for this criterion is the MSE (33)

$$
J\left(\hat{\mathbf{h}}\right) = E\left[e^2(n)\right].
\tag{4.2}
$$

The optimal Wiener filter, $\hat{\mathbf{h}}_W$, is the one that cancels the gradient of $J\left(\hat{\mathbf{h}}\right)$ with respect to $\hat{\mathbf{h}}$, i.e.,

$$
\frac{\partial J\left(\hat{\mathbf{h}}\right)}{\partial \hat{\mathbf{h}}} = \mathbf{0}.
\tag{4.3}
$$

We have

$$\frac{\partial J\left(\hat{\mathbf{h}}\right)}{\partial \hat{\mathbf{h}}} = 2E\left[e(n)\frac{\partial e(n)}{\partial \hat{\mathbf{h}}}\right]$$
$$= -2E\left[e(n)\mathbf{x}(n)\right]. \tag{4.4}$$

Therefore, at the optimum, we have

$$E\left[e_W(n)\mathbf{x}(n)\right] = \mathbf{0}, \tag{4.5}$$

where

$$e_W(n) = d(n) - \hat{\mathbf{h}}_W^T\mathbf{x}(n) \tag{4.6}$$

is the error signal for which $J\left(\hat{\mathbf{h}}\right)$ is minimized (i.e., the optimal filter). Expression (4.5) is called the principle of orthogonality.

The optimal estimate of $y(n)$ is then

$$\hat{y}_W(n) = \hat{\mathbf{h}}_W^T\mathbf{x}(n). \tag{4.7}$$

It is easy to check, with the help of the principle of orthogonality, that we also have

$$E\left[e_W(n)\hat{y}_W(n)\right] = 0. \tag{4.8}$$

The previous expression is called the corollary to the principle of orthogonality.

If we substitute (4.6) into (4.5), we find the Wiener-Hopf equations

$$\mathbf{R_x}\hat{\mathbf{h}}_W = \mathbf{p}_{\mathbf{x}d}, \tag{4.9}$$

where $\mathbf{R_x}$ is the correlation matrix of $\mathbf{x}(n)$ and

$$\mathbf{p}_{\mathbf{x}d} = E\left[\mathbf{x}(n)d(n)\right] \tag{4.10}$$

is the cross-correlation vector between $\mathbf{x}(n)$ and $d(n)$.

The correlation matrix is symmetric and positive semidefinite. It is also Toeplitz, i.e., a matrix which has constant values along diagonals:

$$\mathbf{R_x} = \begin{bmatrix} r_x(0) & r_x(1) & \cdots & r_x(L-1) \\ r_x(1) & r_x(0) & \cdots & r_x(L-2) \\ \vdots & \vdots & \ddots & \vdots \\ r_x(L-1) & r_x(L-2) & \cdots & r_x(0) \end{bmatrix},$$

with $r_x(l) = E[x(n)x(n-l)]$, $l = 0, 1, \ldots, L-1$. For single-channel acoustic and network systems, this matrix is usually positive definite even for signals like speech; however, it can be very ill conditioned.

Assuming that $\mathbf{R_x}$ is non-singular, the optimal Wiener filter is

$$\hat{\mathbf{h}}_W = \mathbf{R_x}^{-1}\mathbf{p}_{xd} \tag{4.11}$$
$$= \mathbf{h}.$$

Solving (4.11) gives exactly the impulse response of the system.

The MSE can be rewritten as

$$J\left(\hat{\mathbf{h}}\right) = \sigma_d^2 - 2\mathbf{p}_{xd}^T\hat{\mathbf{h}} + \hat{\mathbf{h}}^T\mathbf{R_x}\hat{\mathbf{h}}, \tag{4.12}$$

where

$$\sigma_d^2 = E[d^2(n)] \tag{4.13}$$

is the variance of the desired signal $d(n)$. The criterion $J\left(\hat{\mathbf{h}}\right)$ is a quadratic function of the filter coefficient vector $\hat{\mathbf{h}}$ and has a single minimum point. This point combines the optimal Wiener filter, as shown above, and a value called the minimum MSE (MMSE), which is obtained by substituting (4.11) in (4.12):

$$\begin{aligned} J_{\min} &= J\left(\hat{\mathbf{h}}_W\right) \\ &= \sigma_d^2 - \mathbf{p}_{xd}^T\mathbf{R_x}^{-1}\mathbf{p}_{xd} \\ &= \sigma_d^2 - \sigma_{\hat{y}_W}^2, \end{aligned} \tag{4.14}$$

where

$$\sigma_{\hat{y}_W}^2 = E[\hat{y}_W^2(n)] \tag{4.15}$$

is the variance of the optimal filter output signal $\hat{y}_W(n)$. This MMSE can be rewritten as

$$J_{\min} = \sigma_w^2, \tag{4.16}$$

where σ_w^2 is the variance of the noise.

We define the normalized MMSE (NMMSE) as

$$\begin{aligned} \tilde{J}_{\min} &= \frac{J_{\min}}{\sigma_d^2} \\ &= \frac{1}{1+\text{ENR}} \le 1. \end{aligned} \tag{4.17}$$

The previous expression shows how the NMMSE is related to the ENR.

4.1.1   EFFICIENT COMPUTATION OF THE WIENER-HOPF EQUATIONS

In this subsection only, we slightly change the notation in order to make the derivation of an efficient algorithm clearer.

Let

$$\mathbf{x}_L(n) = \begin{bmatrix} x(n) & x(n-1) & \cdots & x(n-L+1) \end{bmatrix}^T$$
$$= \begin{bmatrix} \mathbf{x}_{L-1}^T(n) & x(n-L+1) \end{bmatrix}^T$$

be the far-end signal vector of length L, its corresponding correlation matrix is

$$\mathbf{R}_L = E\left[\mathbf{x}_L(n)\mathbf{x}_L^T(n)\right]$$
$$= \begin{bmatrix} \mathbf{R}_{L-1} & \mathbf{r}_{b,L-1} \\ \mathbf{r}_{b,L-1}^T & r(0) \end{bmatrix}, \tag{4.18}$$

where

$$\mathbf{R}_{L-1} = E[\mathbf{x}_{L-1}(n)\mathbf{x}_{L-1}^T(n)],$$
$$\mathbf{r}_{b,L-1} = \begin{bmatrix} r(L-1) & r(L-2) & \cdots & r(1) \end{bmatrix}^T,$$
$$r(l) = E[x(n)x(n-l)], \ l = 0, 1, \ldots, L-1.$$

The Wiener-Hopf equations are

$$\mathbf{R}_L\hat{\mathbf{h}}_L = \mathbf{p}_L, \tag{4.19}$$

where

$$\mathbf{p}_L = E[\mathbf{x}_L(n)d(n)]$$
$$= \begin{bmatrix} p(0) & p(1) & \cdots & p(L-1) \end{bmatrix}^T$$
$$= \begin{bmatrix} \mathbf{p}_{L-1}^T & E[x(n-L+1)d(n)] \end{bmatrix}^T.$$

We know that

$$\mathbf{R}_{L-1}\mathbf{b}_{L-1} = \mathbf{r}_{b,L-1} \tag{4.20}$$

and

$$\mathbf{R}_L\begin{bmatrix} -\mathbf{b}_{L-1} \\ 1 \end{bmatrix} = \begin{bmatrix} \mathbf{0} \\ E_{L-1} \end{bmatrix}, \tag{4.21}$$

where $\mathbf{b}_{L-1}$ is the backward predictor of length $L-1$ and

$$E_{L-1} = r(0) - \mathbf{r}_{b,L-1}^T\mathbf{b}_{L-1} \tag{4.22}$$

is the prediction error energy. We will use these expressions shortly.

We have

$$\mathbf{R}_L \begin{bmatrix} \hat{\mathbf{h}}_{L-1} \\ 0 \end{bmatrix} = \begin{bmatrix} \mathbf{R}_{L-1} & \mathbf{r}_{b,L-1} \\ \mathbf{r}_{b,L-1}^T & r(0) \end{bmatrix} \begin{bmatrix} \hat{\mathbf{h}}_{L-1} \\ 0 \end{bmatrix}$$

$$= \begin{bmatrix} \mathbf{p}_{L-1} \\ \mathbf{r}_{b,L-1}^T \hat{\mathbf{h}}_{L-1} \end{bmatrix}. \tag{4.23}$$

Using (4.20), the second term of the vector in the right-hand side of (4.23) is also

$$\mathbf{r}_{b,L-1}^T \hat{\mathbf{h}}_{L-1} = \mathbf{b}_{L-1}^T \mathbf{p}_{L-1}. \tag{4.24}$$

Subtracting (4.23) from (4.19), we get

$$\mathbf{R}_L \left\{ \hat{\mathbf{h}}_L - \begin{bmatrix} \hat{\mathbf{h}}_{L-1} \\ 0 \end{bmatrix} \right\} = \begin{bmatrix} \mathbf{0} \\ \varpi_{L-1} \end{bmatrix}, \tag{4.25}$$

where

$$\varpi_{L-1} = p(L-1) - \mathbf{b}_{L-1}^T \mathbf{p}_{L-1}. \tag{4.26}$$

Identifying (4.25) with (4.21), we deduce the recursive equation

$$\hat{\mathbf{h}}_L = \begin{bmatrix} \hat{\mathbf{h}}_{L-1} \\ 0 \end{bmatrix} - \frac{\varpi_{L-1}}{E_{L-1}} \begin{bmatrix} \mathbf{b}_{L-1} \\ -1 \end{bmatrix}. \tag{4.27}$$

The previous expression, along with the Levinson-Durbin algorithm, lead to an efficient way to solve the Wiener-Hopf linear system as shown in Table 4.1. Indeed, it is easy to check that the arithmetic complexity of the algorithm is proportional to L^2. This algorithm is much more efficient than standard methods such as the Gauss elimination technique, whose complexity is in the order of L^3. The other advantage of the Levinson-Durbin algorithm is that it gives the optimal Wiener filters for all orders. Note that in Table 4.1, κ_L is the reflection coefficient and

$$\mathbf{J}_L = \begin{bmatrix} 0 & 0 & \cdots & 0 & 1 \\ 0 & 0 & \cdots & 1 & 0 \\ \vdots & \vdots & \ddots & \vdots & \vdots \\ 0 & 1 & \cdots & 0 & 0 \\ 1 & 0 & \cdots & 0 & 0 \end{bmatrix}.$$

The error signal can also be computed efficiently if we are not interested to find directly the optimal filter. Indeed, if we define the error signal for the filter of order L as

$$e_L(n) = d(n) - \hat{\mathbf{h}}_L^T \mathbf{x}_L(n) \tag{4.28}$$

> **Table 4.1:** Efficient computation of the Wiener-Hopf equations with the Levinson-Durbin algorithm.
>
> | | |
> |---|---|
> | Initialization: | $E_0 = r(0)$ |
> | For | $1 \le l \le L$ |
> | | $\varpi_{l-1} = p(l-1) - \mathbf{b}_{l-1}^T \mathbf{p}_{l-1}$ |
> | | $\kappa_l = \frac{1}{E_{l-1}} \left[r(l) - \mathbf{b}_{l-1}^T \mathbf{J}_{l-1} \mathbf{r}_{b,l-1} \right]$ |
> | | $\hat{\mathbf{h}}_l = \begin{bmatrix} \hat{\mathbf{h}}_{l-1} \\ 0 \end{bmatrix} - \frac{\varpi_{l-1}}{E_{l-1}} \begin{bmatrix} \mathbf{b}_{l-1} \\ -1 \end{bmatrix}$ |
> | | $\mathbf{b}_l = \begin{bmatrix} 0 \\ \mathbf{b}_{l-1} \end{bmatrix} - \kappa_l \mathbf{J}_l \begin{bmatrix} \mathbf{b}_{l-1} \\ -1 \end{bmatrix}$ |
> | | $E_l = E_{l-1} \left(1 - \kappa_l^2 \right)$ |

and plug (4.27) in the previous equation, we easily deduce that

$$e_L(n) = e_{L-1}(n) - \frac{\varpi_{L-1}}{E_{L-1}} e_{b,L-1}(n), \tag{4.29}$$

where

$$e_{b,L-1}(n) = x(n - L + 1) - \mathbf{b}_{L-1}^T \mathbf{x}_{L-1}(n) \tag{4.30}$$

is the backward prediction error signal. It is easy to verify that the error signal of the system can be evaluated without explicitly computing the filters at the different orders.

Solving the Wiener-Hopf equations (4.11) directly or even with the Levinson-Durbin algorithm may not very practical, so adaptive algorithms are usually preferred to find the optimal Wiener filter.

4.2   DETERMINISTIC ALGORITHM

The deterministic or steepest-descent algorithm is actually an iterative algorithm of great importance since it is the starting point of adaptive filters. It is summarized by the simple recursion

$$
\begin{aligned}
\hat{\mathbf{h}}(n) &= \hat{\mathbf{h}}(n-1) - \frac{\mu}{2} \cdot \frac{\partial J\left[\hat{\mathbf{h}}(n-1)\right]}{\partial \hat{\mathbf{h}}(n-1)} \\
&= \hat{\mathbf{h}}(n-1) + \mu \left[\mathbf{p}_{xd} - \mathbf{R}_x \hat{\mathbf{h}}(n-1) \right], \quad n \ge 1, \quad \hat{\mathbf{h}}(0) = \mathbf{0},
\end{aligned}
\tag{4.31}
$$

where μ is a positive constant called the step-size parameter. In this algorithm, $\mathbf{p}_{xd}$ and $\mathbf{R}_x$ are supposed to be known, and clearly, the inversion of the matrix $\mathbf{R}_x$, which can be costly, is not needed. The deterministic algorithm can be reformulated with the error signal:

$$
\begin{aligned}
e(n) &= d(n) - \hat{\mathbf{h}}^T(n-1)\mathbf{x}(n), \tag{4.32} \\
\hat{\mathbf{h}}(n) &= \hat{\mathbf{h}}(n-1) + \mu E[\mathbf{x}(n)e(n)]. \tag{4.33}
\end{aligned}
$$

Now, the important question is: what are the conditions on μ to make the algorithm converge to the true impulse response $\mathbf{h}$? To answer this question, we will examine the natural modes of the algorithm (74).

We define the misalignment vector as

$$\mathbf{m}(n) = \mathbf{h} - \hat{\mathbf{h}}(n), \tag{4.34}$$

which is the difference between the impulse response of the system and the estimated one at iteration n. If we substitute $d(n) = \mathbf{h}^T \mathbf{x}(n) + w(n)$ in the cross-correlation vector, we get

$$\begin{aligned} \mathbf{p}_{\mathbf{x}d} &= E\left[\mathbf{x}(n)d(n)\right] \\ &= \mathbf{R}_{\mathbf{x}}\mathbf{h}. \end{aligned} \tag{4.35}$$

Injecting (4.35) in (4.31) and subtracting $\mathbf{h}$ on both sides of the equation, we obtain

$$\mathbf{m}(n) = (\mathbf{I} - \mu \mathbf{R}_{\mathbf{x}})\mathbf{m}(n-1). \tag{4.36}$$

Using the eigendecomposition of

$$\mathbf{R}_{\mathbf{x}} = \mathbf{Q}\Lambda\mathbf{Q}^T \tag{4.37}$$

in (4.36), where

$$\mathbf{Q}^T\mathbf{Q} = \mathbf{Q}\mathbf{Q}^T = \mathbf{I}, \tag{4.38}$$
$$\Lambda = \operatorname{diag}\left(\lambda_0, \lambda_1, \cdots, \lambda_{L-1}\right), \tag{4.39}$$

and $0 < \lambda_0 \leq \lambda_1 \leq \cdots \leq \lambda_{L-1}$, we get the equivalent form

$$\mathbf{v}(n) = (\mathbf{I} - \mu\Lambda)\mathbf{v}(n-1), \tag{4.40}$$

where

$$\begin{aligned} \mathbf{v}(n) &= \mathbf{Q}^T\mathbf{m}(n) \\ &= \mathbf{Q}^T\left[\mathbf{h} - \hat{\mathbf{h}}(n)\right]. \end{aligned} \tag{4.41}$$

Thus, for the lth natural mode of the steepest-descent algorithm, we have (33)

$$v_l(n) = (1 - \mu\lambda_l)v_l(n-1), \; l = 0, 1, \ldots, L-1, \tag{4.42}$$

or, equivalently,

$$v_l(n) = (1 - \mu\lambda_l)^n v_l(0), \; l = 0, 1, \ldots, L-1. \tag{4.43}$$

The algorithm converges if

$$\lim_{n\to\infty} v_l(n) = 0, \; \forall l. \tag{4.44}$$

In this case

$$\lim_{n \to \infty} \hat{\mathbf{h}}(n) = \mathbf{h}. \tag{4.45}$$

It is straightforward to see from (4.43) that a necessary and sufficient condition for the stability of the deterministic algorithm is that

$$-1 < 1 - \mu\lambda_l < 1, \ \forall l, \tag{4.46}$$

which implies

$$0 < \mu < \frac{2}{\lambda_l}, \ \forall l, \tag{4.47}$$

or

$$0 < \mu < \frac{2}{\lambda_{\max}}, \tag{4.48}$$

where $\lambda_{\max}$ is the largest eigenvalue of the correlation matrix $\mathbf{R_x}$.

Let us evaluate the time needed for each natural mode to converge to a given value. Expression (4.43) gives

$$\ln \frac{|v_l(n)|}{|v_l(0)|} = n \ln |1 - \mu\lambda_l|, \tag{4.49}$$

hence

$$n = \frac{1}{\ln |1 - \mu\lambda_l|} \ln \frac{|v_l(n)|}{|v_l(0)|}. \tag{4.50}$$

The time constant, τ_l for the lth natural mode is defined by taking $|v_l(n)|/|v_l(0)| = 1/e$ (where e is the base of the natural logarithm) in (4.50). Therefore,

$$\tau_l = \frac{-1}{\ln |1 - \mu\lambda_l|}. \tag{4.51}$$

We can link the time constant with the condition number of the correlation matrix $\mathbf{R_x}$. First, let

$$\mu = \frac{\alpha}{\lambda_{\max}}, \tag{4.52}$$

where

$$0 < \alpha < 2, \tag{4.53}$$

to guaranty the convergence of the algorithm, and α is called the normalized step-size parameter. The smallest eigenvalue is $\lambda_{\min} = \lambda_0$; in this case,

$$
\begin{aligned}
\tau_0 &= \frac{-1}{\ln|1 - \alpha\lambda_{\min}/\lambda_{\max}|} \\
&= \frac{-1}{\ln|1 - \alpha/\chi_2[\mathbf{R_x}]|},
\end{aligned}
\tag{4.54}
$$

where $\chi_2[\mathbf{R_x}] = \lambda_{\max}/\lambda_{\min}$ is the condition number of the matrix $\mathbf{R_x}$. We see that the convergence time of the slowest natural mode depends on the conditioning of $\mathbf{R_x}$.

From (4.41), we deduce that

$$
\begin{aligned}
\mathbf{m}^T(n)\mathbf{m}(n) &= \mathbf{v}^T(n)\mathbf{v}(n) \\
&= \left\| \mathbf{h} - \hat{\mathbf{h}}(n) \right\|_2^2 \\
&= \sum_{l=0}^{L-1} \lambda_l(1 - \mu\lambda_l)^n v_l(0).
\end{aligned}
\tag{4.55}
$$

This value gives an idea on the global convergence of the filter to the true impulse response. This convergence is clearly governed by the smallest eigenvalues of $\mathbf{R_x}$.

We now examine the transient behavior of the MSE. Using $d(n) = \mathbf{h}^T\mathbf{x}(n) + w(n)$, the error signal (4.32) can be rewritten as

$$
\begin{aligned}
e(n) &= d(n) - \hat{\mathbf{h}}^T(n-1)\mathbf{x}(n) \\
&= w(n) + \mathbf{m}^T(n-1)\mathbf{x}(n),
\end{aligned}
\tag{4.56}
$$

so that the MSE is

$$
\begin{aligned}
J(n) &= E[e^2(n)] \\
&= \sigma_w^2 + \mathbf{m}^T(n-1)\mathbf{R_x}\mathbf{m}(n-1) \\
&= \sigma_w^2 + \mathbf{v}^T(n-1)\Lambda\mathbf{v}(n-1) \\
&= \sigma_w^2 + \sum_{l=0}^{L-1} \lambda_l(1 - \mu\lambda_l)^{2n-2}v_l^2(0).
\end{aligned}
\tag{4.57}
$$

A plot of $J(n)$ versus n is called the learning curve. Note that the MSE decays exponentially. When the algorithm is convergent, we see that

$$
\lim_{n\to\infty} J(n) = \sigma_w^2.
\tag{4.58}
$$

This value corresponds to the MMSE, $J_{\min}$, obtained with the optimal Wiener filter.

Finally, to end this section, it is worth mentioning that a generalization of the deterministic algorithm is the Newton algorithm

$$
\begin{aligned}
\hat{\mathbf{h}}(n) &= \hat{\mathbf{h}}(n-1) - \left\{ \frac{\partial^2 J\left[\hat{\mathbf{h}}(n-1)\right]}{\partial \hat{\mathbf{h}}^2(n-1)} \right\}^{-1} \frac{\partial J\left[\hat{\mathbf{h}}(n-1)\right]}{\partial \hat{\mathbf{h}}(n-1)} \\
&= \mathbf{R}_{\mathbf{x}}^{-1} \mathbf{p}_{\mathbf{x}d},
\end{aligned}
\tag{4.59}
$$

which converges in one iteration to the optimal Wiener filter. Needless to say that the Newton algorithm is not used at all in echo cancellation.

4.3   STOCHASTIC ALGORITHM

The stochastic gradient or least-mean-square (LMS) algorithm, invented by Widrow and Hoff in the late 50's (73), is certainly the most popular algorithm that we can find in the literature of adaptive filters. The popularity of the LMS is probably due to the fact that it is easy to understand, easy to implement, and robust in many respects.

One convenient way to derive the stochastic gradient algorithm is by approximating the deterministic algorithm. Indeed, in practice, the two quantities $\mathbf{p}_{\mathbf{x}d} = E[\mathbf{x}(n)d(n)]$ and $\mathbf{R}_{\mathbf{x}} = E[\mathbf{x}(n)\mathbf{x}^T(n)]$ are, in general, not known. If we take their instantaneous estimates

$$
\begin{aligned}
\hat{\mathbf{p}}_{\mathbf{x}d}(n) &= \mathbf{x}(n)d(n), \tag{4.60} \\
\hat{\mathbf{R}}_{\mathbf{x}}(n) &= \mathbf{x}(n)\mathbf{x}^T(n), \tag{4.61}
\end{aligned}
$$

and replace them in the steepest-descent algorithm [eq. (4.31)], we get

$$
\begin{aligned}
\hat{\mathbf{h}}(n) &= \hat{\mathbf{h}}(n-1) + \mu\left[\hat{\mathbf{p}}_{\mathbf{x}d}(n) - \hat{\mathbf{R}}_{\mathbf{x}}(n)\hat{\mathbf{h}}(n-1)\right] \\
&= \hat{\mathbf{h}}(n-1) + \mu\mathbf{x}(n)\left[d(n) - \mathbf{x}^T(n)\hat{\mathbf{h}}(n-1)\right]. \tag{4.62}
\end{aligned}
$$

This simple recursion is the LMS algorithm. Contrary to the deterministic algorithm, the LMS weight vector $\hat{\mathbf{h}}(n)$ is now a random vector. The three following equations summarize this algorithm (33):

$$
\begin{aligned}
\hat{y}(n) &= \mathbf{x}^T(n)\hat{\mathbf{h}}(n-1), \text{ filter output}, \tag{4.63} \\
e(n) &= d(n) - \hat{y}(n), \text{ error signal}, \tag{4.64} \\
\hat{\mathbf{h}}(n) &= \hat{\mathbf{h}}(n-1) + \mu\mathbf{x}(n)e(n), \text{ adaptation}, \tag{4.65}
\end{aligned}
$$

which requires $2L$ additions and $2L + 1$ multiplications at each iteration.

The stochastic gradient algorithm has been extensively studied, and many theoretical results on its performance have been obtained (21), (33), (74). In particular, we can show the convergence in the mean and mean square [see for example (75)], where under the independence assumption, the condition is remarkably the same as the one obtained for the deterministic algorithm, i.e.,

$$0 < \mu < \frac{2}{\lambda_{\max}}. \tag{4.66}$$

We can show that the asymptotic MSE for the LMS is

$$\lim_{n \to \infty} J(n) = \sigma_w^2 \left(1 + \frac{\mu}{2} L \sigma_x^2 \right), \tag{4.67}$$

where σ_x^2 is the variance of the signal $x(n)$. If we compare (4.67) with the asymptotic MSE of the steepest-descent algorithm [eq. (4.58)], we notice that a positive term,

$$J_{ex}(\infty) = \frac{\mu}{2} L \sigma_x^2 \sigma_w^2, \tag{4.68}$$

is added, called the excess MSE. This term, of course, has a negative effect on the final MSE, and its apparition is due to the (gradient) approximation discussed at the beginning of this section. We can reduce its effect by taking a very small μ. But taking a small step size will increase the convergence time of the LMS. This tradeoff between fast convergence and increased MSE is a very well-known fact and is something to consider in any practical implementation.

Another interesting definition in adaptive filtering is the misadjustment, which is a measure of how far the steady-state solution computed by the adaptive filter is away from the Wiener solution (33). It is defined as the ratio of $J_{ex}(\infty)$ to $J_{\min}$, i.e.,

$$\mathcal{M} = \frac{J_{ex}(\infty)}{J_{\min}}. \tag{4.69}$$

For the LMS, it is simply

$$\mathcal{M} = \frac{\mu}{2} L \sigma_x^2. \tag{4.70}$$

This misadjustment goes to 0 only if μ goes to 0.

A simple condition for the stability of the LMS is that

$$|\varepsilon(n)| < |e(n)|, \tag{4.71}$$

where

$$\varepsilon(n) = d(n) - \mathbf{x}^T(n)\hat{\mathbf{h}}(n) \tag{4.72}$$

is the a posteriori error signal, computed after the filter is updated. This intuitively makes sense since $\varepsilon(n)$ contains more meaningful information than $e(n)$.

This condition is necessary for the LMS to converge to the impulse response of the system but not sufficient. However, it is very useful to use here and in many other algorithms to find the bounds for the step size μ.

Table 4.2: The normalized LMS (NLMS) algorithm.	
Initialization:	$\hat{\mathbf{h}}(0) = \mathbf{0}$
Parameters:	$0 < \alpha < 2$
	$\delta = \text{cst} \cdot \sigma_x^2$
Error:	$e(n) = d(n) - \mathbf{x}^T(n)\hat{\mathbf{h}}(n-1)$
Update:	$\mu(n) = \dfrac{\alpha}{\mathbf{x}^T(n)\mathbf{x}(n) + \delta}$
	$\hat{\mathbf{h}}(n) = \hat{\mathbf{h}}(n-1) + \mu(n)\mathbf{x}(n)e(n)$

Plugging (4.65) in (4.72) and using the condition (4.71), we find

$$0 < \mu < \frac{2}{\mathbf{x}^T(n)\mathbf{x}(n)}. \tag{4.73}$$

For a large L, $\mathbf{x}^T(n)\mathbf{x}(n) = L\sigma_x^2 = \text{tr}[\mathbf{R_x}]$, where $\text{tr}[\cdot]$ denotes the trace of a square matrix. On the other hand, $\text{tr}[\mathbf{R_x}] = \sum_{l=0}^{L-1} \lambda_l$, and this implies that $\text{tr}[\mathbf{R_x}] \geq \lambda_{\max}$. Hence,

$$0 < \mu < \frac{2}{\mathbf{x}^T(n)\mathbf{x}(n)} \leq \frac{2}{\lambda_{\max}}. \tag{4.74}$$

If we now introduce the normalized step size α ($0 < \alpha < 2$), as we did in the previous section, the step size of the LMS varies with time as follows

$$\mu(n) = \frac{\alpha}{\mathbf{x}^T(n)\mathbf{x}(n)}, \tag{4.75}$$

and the LMS becomes the normalized LMS (NLMS):

$$\hat{\mathbf{h}}(n) = \hat{\mathbf{h}}(n-1) + \frac{\alpha\mathbf{x}(n)e(n)}{\mathbf{x}^T(n)\mathbf{x}(n)}. \tag{4.76}$$

This algorithm is extremely helpful in practice, especially with non-stationary signals, since $\mu(n)$ can adjust itself at each new iteration. In order to avoid numerical difficulties when the energy of the input signal is small, we regularize the algorithm:

$$\hat{\mathbf{h}}(n) = \hat{\mathbf{h}}(n-1) + \frac{\alpha\mathbf{x}(n)e(n)}{\mathbf{x}^T(n)\mathbf{x}(n) + \delta}, \tag{4.77}$$

where $\delta > 0$ is the regularization parameter. Table 4.2 summarizes this very important algorithm. [Note that the definition of $\mu(n)$ in this table is slightly modified, in order to include the regularization parameter δ.]

Since we can approximate $\mu(n)$ by $\alpha/(L\sigma_x^2)$, we can deduce from (4.68) and (4.70) the excess MSE and misadjustment for the NLMS:

$$
\begin{aligned}
J_{\text{ex}}(\infty) &= \frac{\alpha}{2}\sigma_w^2, & (4.78) \\
\mathcal{M} &= \frac{\alpha}{2}. & (4.79)
\end{aligned}
$$

To finish this section, we give another adaptive filter that can be derived in a similar way to the LMS; it is the stochastic Newton algorithm:

$$
\hat{\mathbf{h}}(n) = \hat{\mathbf{h}}(n-1) + \alpha \mathbf{R}_{\mathbf{x}}^{-1}\mathbf{x}(n)e(n). \qquad (4.80)
$$

This algorithm converges much faster than the NLMS but at a very heavy price in terms of numerical complexity.

4.4   VARIABLE STEP-SIZE NLMS ALGORITHM

The stability of the NLMS algorithm is governed by a step-size parameter. As already discussed, the choice of this parameter, within the stability conditions, reflects a tradeoff between fast convergence and good tracking ability on the one hand, and low misadjustment on the other hand. To meet this conflicting requirement, the step size needs to be controlled. While the formulation of this problem is straightforward, a good and reliable solution is not that easy to find. Many different schemes have been proposed in the last two decades: (1), (20), (31), (42), (47), (50), (59), (65). In this section, we show how to derive, in a very simple and elegant way, a non-parametric variable step-size NLMS algorithm.

We can rewrite the a priori and a posteriori error signals as, respectively,

$$
\begin{aligned}
e(n) &= d(n) - \hat{\mathbf{h}}^T(n-1)\mathbf{x}(n) & (4.81) \\
&= \mathbf{x}^T(n)\left[\mathbf{h} - \hat{\mathbf{h}}(n-1)\right] + w(n), \\
\varepsilon(n) &= d(n) - \hat{\mathbf{h}}^T(n)\mathbf{x}(n) & (4.82) \\
&= \mathbf{x}^T(n)\left[\mathbf{h} - \hat{\mathbf{h}}(n)\right] + w(n).
\end{aligned}
$$

Consider the linear update equation:

$$
\hat{\mathbf{h}}(n) = \hat{\mathbf{h}}(n-1) + \mu(n)\mathbf{x}(n)e(n). \qquad (4.83)
$$

One reasonable way to derive a $\mu(n)$ that makes (4.83) stable is to cancel the a posteriori error signal [see (51) and references therein]. Replacing (4.83) in (4.82) with the requirement $\varepsilon(n) = 0$, we easily find assuming $e(n) \neq 0$, $\forall n$, that

$$
\mu_{\text{NLMS}}(n) = \frac{1}{\mathbf{x}^T(n)\mathbf{x}(n)}. \qquad (4.84)
$$

Therefore, the obtained algorithm is the classical NLMS.

While the above procedure makes sense in the absence of noise, finding the $\mu(n)$ in the presence of noise that cancels (4.82) will introduce noise in $\hat{\mathbf{h}}(n)$ since $\mathbf{x}^T(n)\left[\mathbf{h} - \hat{\mathbf{h}}(n)\right] = -w(n) \neq 0, \forall n$. What we would like, in fact, is to have $\mathbf{x}^T(n)\left[\mathbf{h} - \hat{\mathbf{h}}(n)\right] = 0, \forall n$, which implies that $\varepsilon(n) = w(n)$. Hence, in this procedure we wish to find the step-size parameter $\mu(n)$ in such a way that

$$E\left[\varepsilon^2(n)\right] = \sigma_w^2, \forall n. \tag{4.85}$$

Using the approximation $\mathbf{x}^T(n)\mathbf{x}(n) = L\sigma_x^2$ for $L \gg 1$, knowing that $\mu(n)$ is deterministic in nature, substituting (4.83) into (4.82), using (4.81) to eliminate $\hat{\mathbf{h}}(n-1)$, and equating to (4.85), we find

$$
\begin{aligned}
E\left[\varepsilon^2(n)\right] &= \left[1 - \mu(n)L\sigma_x^2\right]^2 \sigma_e^2(n) \\
&= \sigma_w^2,
\end{aligned}
\tag{4.86}
$$

where

$$\sigma_e^2(n) = E\left[e^2(n)\right] \tag{4.87}$$

is the variance of the error signal. Developing (4.86), we obtain a quadratic equation

$$\mu^2(n) - \frac{2}{L\sigma_x^2}\mu(n) + \frac{1}{\left(L\sigma_x^2\right)^2}\left[1 - \frac{\sigma_w^2}{\sigma_e^2(n)}\right] = 0, \tag{4.88}$$

for which the obvious solution is

$$
\begin{aligned}
\mu_{\text{NPVSS}}(n) &= \frac{1}{\mathbf{x}^T(n)\mathbf{x}(n)}\left[1 - \frac{\sigma_w}{\sigma_e(n)}\right] \\
&= \mu_{\text{NLMS}}(n)\alpha_{\text{NPVSS}}(n),
\end{aligned}
\tag{4.89}
$$

where $\alpha_{\text{NPVSS}}(n)$ $[0 \leq \alpha_{\text{NPVSS}}(n) \leq 1]$ is the normalized step size. Therefore, the non-parametric VSS-NLMS (NPVSS-NLMS) algorithm is (9)

$$\hat{\mathbf{h}}(n) = \hat{\mathbf{h}}(n-1) + \mu_{\text{NPVSS}}(n)\mathbf{x}(n)e(n), \tag{4.90}$$

where $\mu_{\text{NPVSS}}(n)$ is defined in (4.89).

We see from (4.89) that before the algorithm converges, $\sigma_e(n)$ is large compared to σ_w, thus $\mu_{\text{NPVSS}}(n) \approx \mu_{\text{NLMS}}(n)$. On the other hand, when the algorithm starts to converge to the true solution, $\sigma_e(n) \approx \sigma_w$ and $\mu_{\text{NPVSS}}(n) \approx 0$. This is exactly what we desire to have both good convergence and low misadjustment. As we can notice, this approach was derived with almost no assumptions compared to all other algorithms belonging to the same family. Table 4.3 summarizes a practical version of the NPVSS-NLMS algorithm.

Table 4.3: The non-parametric VSS-NLMS (NPVSS-NLMS) algorithm.	
Initialization:	$\hat{\mathbf{h}}(0) = \mathbf{0}$ $\hat{\sigma}_e^2(0) = 0$
Parameters:	$\lambda = 1 - \dfrac{1}{KL}$, exponential window with $K \geq 2$ σ_w^2, noise variance known or estimated $\delta = \text{cst} \cdot \sigma_x^2$, regularization $\epsilon > 0$, very small number to avoid division by zero
Error:	$e(n) = d(n) - \hat{\mathbf{h}}^T(n-1)\mathbf{x}(n)$
Update:	$\hat{\sigma}_e^2(n) = \lambda \hat{\sigma}_e^2(n-1) + (1-\lambda)e^2(n)$ $\varsigma(n) = \left[\delta + \mathbf{x}^T(n)\mathbf{x}(n)\right]^{-1}\left[1 - \dfrac{\sigma_w}{\epsilon + \hat{\sigma}_e(n)}\right]$ $\mu_{\text{NPVSS}}(n) = \begin{cases} \varsigma(n) & \text{if } \hat{\sigma}_e(n) \geq \sigma_w \\ 0 & \text{otherwise} \end{cases}$ $\hat{\mathbf{h}}(n) = \hat{\mathbf{h}}(n-1) + \mu_{\text{NPVSS}}(n)\mathbf{x}(n)e(n)$

Similar, at first glance, with the NPVSS-NLMS, a so-called set membership NLMS (SM-NLMS) algorithm was proposed earlier in (28). The step size of this algorithm is

$$\mu_{\text{SM}}(n) = \begin{cases} \mu_{\text{NLMS}}(n)\left[1 - \frac{\eta}{|e(n)|}\right], & \text{if } |e(n)| > \eta \\ \\ 0, & \text{otherwise} \end{cases}, \tag{4.91}$$

where the parameter η represents a bound on the noise. Nevertheless, since there is no averaging on $|e(n)|$, we cannot expect a low misadjustment as for NPVSS-NLMS algorithm. Simulations performed in (9) show that the NPVSS-NLMS outperforms, and by far, the SM-NLMS algorithm, which in fact achieves only a slight performance improvement over the classical NLMS.

4.4.1   CONVERGENCE OF THE MISALIGNMENT

In order to analyze the convergence of the misalignment for the NPVSS-NLMS algorithm, we suppose that the system is stationary. Using the misalignment vector, $\mathbf{m}(n)$, as defined in (4.34), the update equation of the algorithm can be rewritten in terms of the misalignment as

$$\mathbf{m}(n) = \mathbf{m}(n-1) - \mu_{\text{NPVSS}}(n)\mathbf{x}(n)e(n). \tag{4.92}$$

Taking the ℓ_2 norm in (4.92), then the mathematical expectation of both sides, and assuming that

$$E\left[w(n)\mathbf{x}^T(n)\mathbf{m}(n-1)\right] = 0, \tag{4.93}$$

which is true if $w(n)$ is white, we obtain

$$
\begin{aligned}
E\left[\|\mathbf{m}(n)\|_2^2\right] - E\left[\|\mathbf{m}(n-1)\|_2^2\right] &= -\mu_{\text{NPVSS}}(n)\left[\sigma_e(n) - \sigma_w\right]\left[\sigma_e(n) + 2\sigma_w\right] \\
&\leq 0.
\end{aligned}
\tag{4.94}
$$

The previous expression proves that the length of the misalignment vector for the NPVSS-NLMS algorithm is nonincreasing, which implies that

$$
\lim_{n \to \infty} \sigma_e^2(n) = \sigma_w^2.
\tag{4.95}
$$

It should be noticed that the previous relation does not imply that $E\left[\|\mathbf{m}(\infty)\|_2^2\right] = 0$. However, under the independence assumption, we can show the equivalence. Indeed, from

$$
e(n) = \mathbf{x}^T(n)\mathbf{m}(n-1) + w(n),
\tag{4.96}
$$

it can be shown that

$$
E\left[e^2(n)\right] = \sigma_w^2 + \text{tr}\left[\mathbf{R_x}\mathbf{K}(n-1)\right]
\tag{4.97}
$$

if $\mathbf{x}(n)$ are independent, where $\mathbf{K}(n-1) = E\left[\mathbf{m}(n-1)\mathbf{m}^T(n-1)\right]$. Taking (4.95) into account, (4.97) becomes

$$
\text{tr}\left[\mathbf{R_x}\mathbf{K}(\infty)\right] = 0.
\tag{4.98}
$$

Assuming that $\mathbf{R_x} > 0$, it results that $\mathbf{K}(\infty) = \mathbf{0}$, and consequently

$$
E\left[\|\mathbf{m}(\infty)\|_2^2\right] = 0.
\tag{4.99}
$$

So, in principle, the excess MSE and misadjustment for the NPVSS-NLMS should be

$$
\begin{aligned}
J_{\text{ex}}(\infty) &= 0, \tag{4.100}\\
\mathcal{M} &= 0. \tag{4.101}
\end{aligned}
$$

4.5   SIGN ALGORITHMS

Up to now, the only cost function that we have used is the MSE. What makes this criterion so interesting is that an optimal solution (Wiener) can be easily derived as well as very powerful adaptive algorithms. An alternative to the MSE is the mean absolute error (MAE) (27):

$$
\begin{aligned}
J_a\left(\hat{\mathbf{h}}\right) &= E\left[|e(n)|\right] \tag{4.102}\\
&= E\left[\left|d(n) - \hat{\mathbf{h}}^T\mathbf{x}(n)\right|\right].
\end{aligned}
$$

The gradient of this cost function is

$$\frac{\partial J_a\left(\hat{\mathbf{h}}\right)}{\partial \hat{\mathbf{h}}} = -E\left\{\mathbf{x}(n)\text{sgn}[e(n)]\right\}, \tag{4.103}$$

where

$$\text{sgn}[e(n)] = \frac{e(n)}{|e(n)|}. \tag{4.104}$$

From the instantaneous value of the gradient of $J_a\left(\hat{\mathbf{h}}\right)$, we can derive the sign-error adaptive filter:

$$\hat{\mathbf{h}}(n) = \hat{\mathbf{h}}(n-1) + \mu_a\mathbf{x}(n)\text{sgn}[e(n)], \tag{4.105}$$

where μ_a is the adaptation step of the algorithm. This algorithm is simplified, compared to the LMS, since the L multiplications in the update equation are replaced by a sign change of the components of the signal vector $\mathbf{x}(n)$. Using the stability condition, $|\varepsilon(n)| < |e(n)|$, we deduce that

$$0 < \mu_a < \frac{2|e(n)|}{\mathbf{x}^T(n)\mathbf{x}(n)}. \tag{4.106}$$

Another way to simplify the LMS filter is to replace $x(n)$ with its sign. We get the signed regressor LMS (SR-LMS) algorithm (12):

$$\hat{\mathbf{h}}(n) = \hat{\mathbf{h}}(n-1) + \mu_a'\text{sgn}[\mathbf{x}(n)]e(n), \tag{4.107}$$

where μ_a' is the adaptation step of the algorithm and the stability condition is

$$0 < \mu_a' < \frac{2}{\mathbf{x}^T(n)\text{sgn}[\mathbf{x}(n)]}. \tag{4.108}$$

Combining the two previous approaches, we derive the sign-sign algorithm:

$$\hat{\mathbf{h}}(n) = \hat{\mathbf{h}}(n-1) + \mu_a''\text{sgn}[\mathbf{x}(n)]\text{sgn}[e(n)], \tag{4.109}$$

for which the stability condition is

$$0 < \mu_a'' < \frac{2|e(n)|}{\mathbf{x}^T(n)\text{sgn}[\mathbf{x}(n)]}. \tag{4.110}$$

The algorithms derived in this section are very simple to implement and can be very useful in echo cancellation. However, their convergence rate is usually lower than that of the LMS and their excess MSE is higher (2), (10), (11).

CHAPTER 5

Basic Proportionate-Type NLMS Adaptive Filters

In telephone networks, it has been known for several decades that a typical echo path impulse response is sparse. The echo part itself is usually short, but the bulk delay can be much longer and can also vary significantly. As a result, a long adaptive filter is needed to cover both the echo part and the delay. Therefore, only a small number of the coefficients of the adaptive filter are really active (i.e., different from zero), and all others are inactive (i.e., small values or zero).

It was desirable for a long time to take advantage of the sparsity of the network impulse response in order to improve the convergence rate and tracking of the NLMS algorithm as it is well known that it converges slowly for large filter lengths. Duttweiler was one of the first researchers to come up with an elegant idea, more than a decade ago, by proposing the proportionate NLMS (PNLMS) algorithm (18). The idea behind the PNLMS is to update each coefficient of the filter independently of the others, by adjusting the adaptation step size in proportion to the magnitude of the estimated filter coefficient. It redistributes the adaptation gains among all coefficients and emphasizes the large ones (in magnitude) in order to speed up their convergence and, consequently, achieving a fast initial convergence rate. Since then, many other sophisticated algorithms working on the same principle have been proposed with the ultimate purpose to fully exploit sparsity in network and acoustic impulse responses.

This chapter reviews the basic proportionate-type NLMS adaptive filters.

5.1 GENERAL DERIVATION

In this section, we show how to derive any proportionate-type NLMS algorithm by using a simplified version of the universal criterion proposed in (38).

As explained in (41), a reasonable adaptive algorithm must find a good balance between its needs to be conservative (retain the information it has acquired in preceding iterations) and corrective (make sure that with new information, the accuracy of the solution is increased). Therefore, a good criterion should be the sum of two terms reflecting mathematically this principle. One of these terms is simply a distance between the old and new weight vectors (and depending on how we define this distance, we obtain different update rules), and the other one depends on the a posteriori error signal. According to this principle, one easy way to find proportionate-type NLMS adaptive filters that adjust the new weight vector, $\hat{\mathbf{h}}(n)$, from the old one, $\hat{\mathbf{h}}(n-1)$, is to minimize the following cost function:

$$
\begin{aligned}
J(n) &= D\left[\hat{\mathbf{h}}(n), \hat{\mathbf{h}}(n-1)\right] + \varepsilon^2(n) \\
&= D\left[\hat{\mathbf{h}}(n), \hat{\mathbf{h}}(n-1)\right] + \left[d(n) - \mathbf{x}^T(n)\hat{\mathbf{h}}(n)\right]^2,
\end{aligned}
\tag{5.1}
$$

where $D\left[\hat{\mathbf{h}}(n), \hat{\mathbf{h}}(n-1)\right]$ is some measure of distance from the old to the new weight vectors. Differentiating $J(n)$ with respect to $\hat{\mathbf{h}}(n)$ and setting the resulting vector to zero, we can see that the adaptive algorithm has the form:

$$
\mathbf{x}(n)\mathbf{x}^T(n)\left[\hat{\mathbf{h}}(n) - \hat{\mathbf{h}}(n-1)\right] + \frac{1}{2}\frac{\partial D\left[\hat{\mathbf{h}}(n), \hat{\mathbf{h}}(n-1)\right]}{\partial \hat{\mathbf{h}}(n)} = \mathbf{x}(n)e(n).
\tag{5.2}
$$

We choose for the distance

$$
D\left[\hat{\mathbf{h}}(n), \hat{\mathbf{h}}(n-1)\right] = \delta\left[\hat{\mathbf{h}}(n) - \hat{\mathbf{h}}(n-1)\right]^T \mathbf{Q}^{-1}(n)\left[\hat{\mathbf{h}}(n) - \hat{\mathbf{h}}(n-1)\right],
\tag{5.3}
$$

where $\mathbf{Q}(n)$ is a positive-definite symmetric matrix and $\delta > 0$ is the regularization parameter. Using (5.3) in (5.2), we obtain

$$
\left[\delta\mathbf{Q}^{-1}(n) + \mathbf{x}(n)\mathbf{x}^T(n)\right]\left[\hat{\mathbf{h}}(n) - \hat{\mathbf{h}}(n-1)\right] = \mathbf{x}(n)e(n),
\tag{5.4}
$$

which implies that

$$
\mathbf{Q}^{-1/2}(n)\left[\hat{\mathbf{h}}(n) - \hat{\mathbf{h}}(n-1)\right] = \left[\delta\mathbf{I} + \mathbf{Q}^{1/2}(n)\mathbf{x}(n)\mathbf{x}^T(n)\mathbf{Q}^{1/2}(n)\right]^{-1}\mathbf{Q}^{1/2}(n)\mathbf{x}(n)e(n).
\tag{5.5}
$$

Using the fact that

$$
\left[\delta\mathbf{I} + \mathbf{Q}^{1/2}(n)\mathbf{x}(n)\mathbf{x}^T(n)\mathbf{Q}^{1/2}(n)\right]^{-1}\mathbf{Q}^{1/2}(n)\mathbf{x}(n) = \frac{\mathbf{Q}^{1/2}(n)\mathbf{x}(n)}{\delta + \mathbf{x}^T(n)\mathbf{Q}(n)\mathbf{x}(n)},
\tag{5.6}
$$

we get the update equation:

$$
\hat{\mathbf{h}}(n) = \hat{\mathbf{h}}(n-1) + \frac{\mathbf{Q}(n)\mathbf{x}(n)e(n)}{\delta + \mathbf{x}^T(n)\mathbf{Q}(n)\mathbf{x}(n)}.
\tag{5.7}
$$

To better control the convergence rate and misadjustment of the algorithm, we can also add in (5.7) the normalized step-size factor α to finally obtain the general form of the adaptive filter:

$$
\hat{\mathbf{h}}(n) = \hat{\mathbf{h}}(n-1) + \frac{\alpha\mathbf{Q}(n)\mathbf{x}(n)e(n)}{\delta + \mathbf{x}^T(n)\mathbf{Q}(n)\mathbf{x}(n)}.
\tag{5.8}
$$

Taking $\mathbf{Q}(n) = \mathbf{I}$ in (5.8), we find the NLMS algorithm.

Now let us assume, for an instant, that δ is negligible or zero. From the stability condition, $|\varepsilon(n)| < |e(n)|$, we deduce that $0 < \alpha < 2$. Therefore, as long as the symmetric matrix $\mathbf{Q}(n)$ is positive definite and $0 < \alpha < 2$, the adaptive filter in (5.8) is guarantied to not diverge.

In the rest of this work we choose

$$\mathbf{Q}(n) = \mathbf{G}(n - 1), \tag{5.9}$$

where

$$\mathbf{G}(n - 1) = \mathrm{diag}\begin{bmatrix} g_0(n - 1) & g_1(n - 1) & \cdots & g_{L-1}(n - 1) \end{bmatrix} \tag{5.10}$$

is an $L \times L$ diagonal matrix, with $g_l(n - 1) > 0$, $\forall n, l$. Furthermore, $\mathbf{G}(n - 1)$ will depend on $\hat{\mathbf{h}}(n - 1)$; that is what characterizes proportionate-type NLMS adaptive filters. Depending on how the elements $g_l(n - 1)$ are chosen, we obtain different kind of proportionate algorithms. We will also normalize the diagonal elements of $\mathbf{G}(n - 1)$ as follows:

$$g_l(n - 1) = \frac{\gamma_l(n - 1)}{\sum_{i=0}^{L-1} \gamma_i(n - 1)}, \tag{5.11}$$

to avoid any strange problems during regularization of the algorithms.

5.2 THE PROPORTIONATE NLMS (PNLMS) AND PNLMS++ ALGORITHMS

As explained at the beginning of this chapter, it is important for an adaptive filter to rapidly emphasize on the identification of the active coefficients in sparse impulse responses. An algorithm that does that pretty well is the proportionate NLMS (PNLMS), proposed by Duttweiler (18). In this algorithm, an adaptive individual step size is assigned to each filter coefficient. The step sizes are calculated from the last estimate of the filter coefficients, so that larger coefficients receive a larger step size, thus increasing the convergence rate of that coefficient. This has the effect that active coefficients are adjusted faster than non-active coefficients (i.e., small or zero coefficients). An advantage of this technique compared to other approaches to individual step-size algorithms, e.g., (35), (69), is that less a priori information is needed. Furthermore, the active regions of the echo path do not have to be detected explicitly. In the PNLMS, the diagonal elements of $\mathbf{G}(n - 1)$ are calculated as (18)

$$\gamma_{\min}(n - 1) = \varrho \max\left[\delta_{\mathrm{p}}, \left|\hat{h}_0(n - 1)\right|, \ldots, \left|\hat{h}_{L-1}(n - 1)\right| \right], \tag{5.12a}$$

$$\gamma_l(n - 1) = \max\left[\gamma_{\min}(n - 1), \left|\hat{h}_l(n - 1)\right| \right], \quad 0 \leq l \leq L - 1, \tag{5.12b}$$

where the parameters δ_{p} and ϱ are positive numbers with typical values $\delta_{\mathrm{p}} = 0.01$, $\varrho = 5/L$. The term $\gamma_{\min}(n - 1)$ prevents $\left|\hat{h}_l(n - 1)\right|$ from stalling when it is much smaller than the magnitude of

Table 5.1: The proportionate NLMS (PNLMS) algorithm.

Initialization:	$\hat{h}_l(0) = 0, \; l = 0, 1, \ldots, L - 1$
Parameters:	$\delta_p = 0.01$
	$\varrho = 5/L$
	$0 < \alpha < 2$
	$\delta_{PNLMS} = \text{cst} \cdot \sigma_x^2 / L$
Error:	$e(n) = d(n) - \mathbf{x}^T(n)\hat{\mathbf{h}}(n - 1)$
Update:	$\gamma_{\min}(n - 1) = \varrho \max\left[\delta_p, \left|\hat{h}_0(n - 1)\right|, \ldots, \left|\hat{h}_{L-1}(n - 1)\right|\right]$
	$\gamma_l(n - 1) = \max\left[\gamma_{\min}(n - 1), \left|\hat{h}_l(n - 1)\right|\right]$
	$g_l(n - 1) = \dfrac{\gamma_l(n - 1)}{\sum_{i=0}^{L-1} \gamma_i(n - 1)}, \; l = 0, 1, \ldots, L - 1$
	$\mu(n) = \dfrac{\alpha}{\sum_{i=0}^{L-1} x^2(n - i)g_i(n - 1) + \delta_{PNLMS}}$
	$\hat{h}_l(n) = \hat{h}_l(n - 1) + \mu(n)g_l(n - 1)x(n - l)e(n),$
	$l = 0, 1, \ldots, L - 1$

the largest coefficient, and δ_p regularizes the update when all coefficients are zero at initialization. For the regularization parameter, we usually choose

$$\delta_{PNLMS} = \frac{\delta_{NLMS}}{L}, \tag{5.13}$$

where δ_{NLMS} is the regularization parameter for the NLMS.

Table 5.1 summarizes the PNLMS algorithm.

A variant of this algorithm is called the PNLMS++ (25). In this algorithm, for odd-numbered time steps, the matrix $\mathbf{G}(n - 1)$ is derived as above, while for even-numbered steps, it is chosen to be the identity matrix, i.e.,

$$\mathbf{G}(n - 1) = \mathbf{I}, \tag{5.14}$$

which leads to an NLMS iteration. The alternation between NLMS and PNLMS iterations has advantages, compared to using just the PNLMS technique, e.g., it makes the PNLMS++ algorithm much less sensitive to the assumption of a sparse impulse response without sacrificing performance.

5.3   THE SIGNED REGRESSOR PNLMS ALGORITHM

The PNLMS algorithm requires roughly $4L$ multiplications at each iteration. It is the double of the NLMS complexity. One way to reduce this complexity is to combine the proportionate idea with the SR-LMS algorithm as proposed in (5). We recall that the performance of the SR-LMS is similar to the LMS while its number of multiplications is two times smaller.

For a nonstationary signal like speech, the SR-LMS is not properly normalized, i.e., it has a different convergence rate, depending on the power of the excitation signal. To alleviate this problem, we use the normalized version proposed in (5) where the update is divided by the sum of the magnitude of the input samples, giving

$$\hat{\mathbf{h}}(n) = \hat{\mathbf{h}}(n-1) + \frac{\alpha \, \text{sign}\,[\mathbf{x}(n)]\, e(n)}{\mathbf{x}^T(n)\text{sign}\,[\mathbf{x}(n)] + \delta/\left\{\mathbf{x}^T(n)\text{sign}\,[\mathbf{x}(n)] + \delta_r\right\}}. \tag{5.15}$$

This normalization is reasonable since $\text{sign}\,[\mathbf{x}(n)]\, e(n) = \text{sign}\,[\mathbf{x}(n)]\left\{\left[\mathbf{h} - \hat{\mathbf{h}}(n)\right]^T \mathbf{x}(n) + w(n)\right\}$ is proportional to the magnitude of $x(n)$ for $w(n)$ small. Furthermore, the regularization parameter δ is also scaled by $\sum_{l=0}^{L-1}|x(n-l)| + \delta_r = \mathbf{x}^T(n)\text{sign}\,[\mathbf{x}(n)] + \delta_r$, where δ_r inhibits division by zero in (5.15). With this scaling of δ, the effect of regularizing (5.15) is the same as regularizing with the same δ in the NLMS.

We now propose to combine the properly normalized and regularized signed regressor algorithm with the proportionate step-size matrix $\mathbf{G}(n-1)$. Our motivation for doing this is to remedy the reduced convergence rate that results from using the SR-LMS, compared to NLMS, and achieve a reduction of the complexity compared to PNLMS. We start with the basic signed regressor PNLMS (SR-PNLMS) algorithm,

$$\hat{\mathbf{h}}(n) = \hat{\mathbf{h}}(n-1) + \frac{\alpha \mathbf{G}(n-1)\text{sign}\,[\mathbf{x}(n)]\, e(n)}{\mathbf{x}^T(n)\mathbf{G}(n-1)\text{sign}\,[\mathbf{x}(n)] + \delta/\left\{\mathbf{x}^T(n)\text{sign}\,[\mathbf{x}(n)] + \delta_r\right\}}. \tag{5.16}$$

A "++" version, in the same manner as for PNLMS++, can also be given for this algorithm by alternating between (5.16) and (5.15).

Algorithm (5.16) still has a rather high numerical complexity because of the term, $\mathbf{x}^T(n)\mathbf{G}(n-1)\text{sign}\,[\mathbf{x}(n)]$, which consumes L multiplications. A more efficient normalization that works well for sparse echo paths is

$$\hat{\mathbf{h}}(n) = \hat{\mathbf{h}}(n-1) + \frac{\alpha \mathbf{G}(n-1)\text{sign}\,[\mathbf{x}(n)]\, e(n)}{\theta g_{\max}(n-1)\mathbf{x}^T(n)\text{sign}\,[\mathbf{x}(n)] + \delta/\left\{\mathbf{x}^T(n)\text{sign}\,[\mathbf{x}(n)] + \delta_r\right\}}, \tag{5.17}$$

where

$$g_{\max}(n-1) = \max_l g_l(n-1) \tag{5.18}$$

and θ is a constant, which is adjusted to make (5.17) behave similar to (5.16). Typically, $\theta = 10/L$. For dispersive echo paths however, this algorithm may become unstable. Note that this technique for reducing complexity can be used in the PNLMS algorithm as well.

5.4   THE IMPROVED PNLMS (IPNLMS) ALGORITHMS

When the echo path is sparse, the PNLMS has a very fast initial convergence and a good initial tracking. Unfortunately, when the impulse response is dense, the PNLMS converges much more slowly than the NLMS. This implies that the rule proposed in PNLMS is far from optimal. The intuitive reason behind this deterioration is that the proportionality rule in the PNLMS may be too aggressive.

5.4.1   THE REGULAR IPNLMS

The idea behind the improved PNLMS (IPNLMS) algorithm is to try to attenuate the aggressiveness of the PNLMS by combining the NLMS update and a proportionate term in the same rule. In the IPNLMS (6), we choose

$$\gamma_l(n-1) = (1-\kappa)\frac{\left\|\hat{\mathbf{h}}(n-1)\right\|_1}{L} + (1+\kappa)\left|\hat{h}_l(n-1)\right|, \ 0 \le l \le L-1, \tag{5.19}$$

therefore

$$
\begin{aligned}
g_l(n-1) &= \frac{\gamma_l(n-1)}{\sum_{i=0}^{L-1}\gamma_i(n-1)} \\
&= \frac{1-\kappa}{2L} + (1+\kappa)\frac{\left|\hat{h}_l(n-1)\right|}{2\left\|\hat{\mathbf{h}}(n-1)\right\|_1}, \ 0 \le l \le L-1,
\end{aligned}
\tag{5.20}
$$

where κ ($-1 \le \kappa < 1$) is a parameter that controls the amount of proportionality in the IPNLMS. For $\kappa = -1$, it can be easily checked that the IPNLMS and NLMS algorithms are identical. For κ close to 1, IPNLMS behaves like PNLMS. In practice, a good choice for κ is -0.5 or 0. With this choice and in simulations, IPNLMS always performs better than NLMS and PNLMS.

Now, we briefly discuss the regularization of the IPNLMS. At initialization, since all the taps of the filter start with zero, the vector $\mathbf{x}(0)$ is multiplied by $(1-\kappa)/(2L)$. This suggests that the regularization of the IPNLMS algorithm should be taken as

$$\delta_{\text{IPNLMS}} = \frac{1-\kappa}{2L}\delta_{\text{NLMS}}. \tag{5.21}$$

The IPNLMS adaptive filter is summarized in Table 5.2.

One important parameter in the IPNLMS is κ. We can make it variable to try to improve the performance of the IPNLMS. Ideally, when the echo path is very sparse, it is desirable to have a value of κ close to 1; however, when the echo path is dense, a value of κ close to -1 is more appropriate. In (43), an interesting idea along those lines was proposed.

First, let us make the change of variable[1]

$$\kappa'(n-1) = \frac{1+\kappa}{2} \tag{5.22}$$

---

[1]Note that now we make the new variable time dependent.

Table 5.2: The improved PNLMS (IPNLMS) algorithm.

Initialization:	$\hat{h}_l(0) = 0,\ l = 0, 1, \dots, L - 1$
Parameters:	$-1 \le \kappa < 1$
	$0 < \alpha < 2$
	$\delta_{\text{IPNLMS}} = \text{cst} \cdot \sigma_x^2 \dfrac{1 - \kappa}{2L}$
	$\epsilon > 0$, very small number to avoid division by zero
Error:	$e(n) = d(n) - \mathbf{x}^T(n)\hat{\mathbf{h}}(n - 1)$
Update:	$g_l(n - 1) = \dfrac{1 - \kappa}{2L} + (1 + \kappa)\dfrac{\left|\hat{h}_l(n - 1)\right|}{2\left\|\hat{\mathbf{h}}(n - 1)\right\|_1 + \epsilon},$
	$l = 0, 1, \dots, L - 1$
	$\mu(n) = \dfrac{\alpha}{\sum_{i=0}^{L-1} x^2(n - i)g_i(n - 1) + \delta_{\text{IPNLMS}}}$
	$\hat{h}_l(n) = \hat{h}_l(n - 1) + \mu(n)g_l(n - 1)x(n - l)e(n),$
	$l = 0, 1, \dots, L - 1$

in (5.20). We obtain

$$g_l(n - 1) = \frac{1 - \kappa'(n - 1)}{L} + \kappa'(n - 1)\frac{\left|\hat{h}_l(n - 1)\right|}{\left\|\hat{\mathbf{h}}(n - 1)\right\|_1},\ 0 \le l \le L - 1. \tag{5.23}$$

This time we have $0 \le \kappa'(n - 1) < 1$. Thus, when $\kappa'(n - 1) = 0$ we get the NLMS, and when $\kappa'(n - 1)$ is close to 1, we get a behavior close to the PNLMS. We can now link $\kappa'(n - 1)$ with any of the sparseness measures (say ξ_{12}) presented in Chapter 2 since both variables vary from 0 to 1 and go in the same direction, i.e., when the impulse response is very sparse, ξ_{12} is close to 1 and a value of $\kappa'(n - 1)$, close to 1 is desirable; on the other hand, when the impulse response is dense, ξ_{12} is close to 0 and a value of $\kappa'(n - 1)$, close to 0 is preferable. Therefore, for the IPNLMS to behave very well, the variable $\kappa'(n - 1)$ should follow the sparseness measure ξ_{12}. An experimental study in (43) shows that

- for a sparse impulse response ($\xi_{12} > 0.6$), a value for $\kappa'(n - 1)$ of 0.6 or larger is a good choice;

- for a non-sparse impulse response ($0.2 \le \xi_{12} \le 0.6$), a $\kappa'(n - 1)$ between 0.1 and 0.2 is appropriate; and

- for a dense impulse response ($\xi_{12} < 0.2$), $\kappa'(n - 1)$ should be smaller than 0.2.

The authors from this experimental study propose then to link $\kappa'(n - 1)$ and ξ_{12} with a curve fitting of a cubic function:

$$\kappa'(n - 1) = 3\xi_{12}^3\left[\hat{\mathbf{h}}(n - 1)\right] - 3\xi_{12}^2\left[\hat{\mathbf{h}}(n - 1)\right] + \xi_{12}\left[\hat{\mathbf{h}}(n - 1)\right]. \tag{5.24}$$

Simulations in (43) show that the IPNLMS with the variable $\kappa'(n-1)$ gives very good performances.

Another idea with a variable parameter in the IPNLMS, similar to the one just presented, can be found in (45).

5.4.2   THE IPNLMS WITH THE ℓ_0 NORM

It can be noticed that the regular IPNLMS algorithm uses the ℓ_1 norm to exploit the sparsity of the impulse response that we need to identify. A better measure can be the ℓ_0 norm since it is a natural mathematical measure of sparseness (16), (52). However, the function $f(h_l)$, associated with the ℓ_0 norm (see Chapter 2), is not continuous, and because that many elements of the vector $\mathbf{h}$ can be very small but not exactly zero, it is better to approximate it by a smooth and continuous function. A good approximation is (30)

$$f(h_l) \approx 1 - e^{-\beta_0|h_l|}, \tag{5.25}$$

where β_0 is a large positive value. Therefore,

$$
\begin{aligned}
\|\mathbf{h}\|_0 &= \lim_{\beta_0 \to \infty} \sum_{l=0}^{L-1} \left[1 - e^{-\beta_0|h_l|} \right] \\
&\approx \sum_{l=0}^{L-1} \left[1 - e^{-\beta_0|h_l|} \right]
\end{aligned}
\tag{5.26}
$$

for a large positive β_0. Now we can use this norm to estimate the elements of $\mathbf{G}(n-1)$. Following the principle of the regular IPNLMS, we have

$$\gamma_l(n-1) = (1-\kappa)\frac{\left\|\hat{\mathbf{h}}(n-1)\right\|_0}{L} + (1+\kappa)\left[1 - e^{-\beta_0\left|\hat{h}_l(n-1)\right|}\right], \ 0 \leq l \leq L-1. \tag{5.27}$$

Next, substituting (5.27) into (5.11) and taking (5.26) into account, we find that (57)

$$g_l(n-1) = \frac{1-\kappa}{2L} + (1+\kappa)\frac{\left[1 - e^{-\beta_0\left|\hat{h}_l(n-1)\right|}\right]}{2\left\|\hat{\mathbf{h}}(n-1)\right\|_0}, \ 0 \leq l \leq L-1. \tag{5.28}$$

Some practical issues should be outlined. First, the choice of the parameter β_0 is important. It should depend on the sparseness of the impulse response. For a dense echo path, a large value for β_0 should be required. For a very sparse echo path, a small value for β_0 is necessary. Several considerations regarding the choice of this parameter can be found in (30). Also, we can obtain some a priori information about the impulse response by using the regular IPNLMS in the first iterations, estimate its sparsity with a sparseness measure, and then choose the value for β_0 accordingly.

Second, the evaluation of the exponential term in (5.28) could be problematic in practical implementations. A possible solution is based on the first order Taylor series expansions of exponential functions (30). Another practical option is to use a look-up table.

Simulations in (57) show that the IPNLMS with the ℓ_0 norm performs better than the regular IPNLMS when the echo path is very sparse.

To finish this subsection, let us show in a very intuitive way how much the convergence rate of a proportionate-type NLMS algorithm can potentially be improved as compared to the NLMS algorithm. Let us assume that the echo path has exactly L_a active coefficients and all others $(L - L_a)$ are exactly zero. Taking the limiting case $\kappa = 1$ in the IPNLMS with the ℓ_0 norm, we get

$$g_l(n - 1) = \frac{f\left[\hat{h}_l(n - 1)\right]}{L_a}, \ 0 \le l \le L - 1. \tag{5.29}$$

If we further assume that the far-end signal is stationary and neglect the effect of the regularization parameter, we easily deduce the update equation for the IPNLMS:

$$\hat{\mathbf{h}}(n) = \hat{\mathbf{h}}(n - 1) + \frac{\alpha}{L_a} \cdot \frac{\mathbf{G}_{01}(n - 1)\mathbf{x}(n)e(n)}{\sigma_x^2}, \tag{5.30}$$

where $\mathbf{G}_{01}(n - 1)$ is a diagonal matrix containing only 0's and 1's. In the same conditions, the update equation for the NLMS is

$$\hat{\mathbf{h}}(n) = \hat{\mathbf{h}}(n - 1) + \frac{\alpha}{L} \cdot \frac{\mathbf{x}(n)e(n)}{\sigma_x^2}. \tag{5.31}$$

Comparing (5.30) with (5.31), we see that the convergence rate of the IPNLMS can be improved by a factor as large as L/L_a with some a priori information. Ideally, the IPNLMS is the NLMS with a filter length equal to L_a; and as we know, the shorter the NLMS filter is the faster it converges. For example, with an echo path of length $L = 500$ which has only $L_a = 50$ active taps, the convergence rate of the IPNLMS can be 10 times faster than the NLMS with the same misadjustment. However, with real impulse responses, we are still far away from this upper bound.

5.4.3   THE IPNLMS WITH A NORM-LIKE DIVERSITY MEASURE

The p-norm-like, $\ell_{(0 \le p \le 1)}$, diversity measure is defined as (40), (60)

$$\mathcal{D}_p(\mathbf{h}) = \sum_{l=0}^{L-1} |h_l|^p, \ 0 \le p \le 1. \tag{5.32}$$

It is clear that for $p = 0$, we get the ℓ_0 norm and for $p = 1$, we get the ℓ_1 norm. However, $\ell_{(0<p<1)}$ is not a true norm (40).

Using the previous diversity measure in the IPNLMS, we obtain:

$$\gamma_l(n - 1) = (1 - \kappa)\frac{\mathcal{D}_p\left[\hat{\mathbf{h}}(n - 1)\right]}{L} + (1 + \kappa)\left|\hat{h}_l(n - 1)\right|^p, \ 0 \le l \le L - 1. \tag{5.33}$$

As a result, the diagonal elements of the matrix $\mathbf{G}(n-1)$ are

$$g_l(n-1) = \frac{1-\kappa}{2L} + (1+\kappa)\frac{\left|\hat{h}_l(n-1)\right|^p}{2\mathcal{D}_p\left[\hat{\mathbf{h}}(n-1)\right]}, \quad 0 \le l \le L-1. \tag{5.34}$$

Obviously, the IPNLMS with the ℓ_0 and ℓ_1 norms are particular cases of the IPNLMS using the diversity measure.

CHAPTER 6

The Exponentiated Gradient Algorithms

Around the same time when the PNLMS was invented, another variant of the LMS algorithm, called the exponentiated gradient algorithm with positive and negative weights (EG± algorithm), was proposed by Kivinen and Warmuth (41). This algorithm converges much faster than the LMS algorithm when the impulse response that we need to identify is sparse, which is often the case in network echo cancellation involving a hybrid transformer in conjunction with variable network delay, or in the context of hands-free communications where there is a strong coupling between the loudspeaker and the microphone (5). The EG± algorithm has the nice feature that its update rule takes advantage of the sparseness of the impulse response to speed up its initial convergence and to improve its tracking abilities as compared to the LMS. In (34), a general expression of the MSE is derived for the EG± algorithm showing that for sparse impulse responses, the EG± algorithm, like PNLMS, converges more quickly than the LMS for a given asymptotic MSE.

In this chapter, we show how to derive the EG algorithms and demonstrate that they are connected to some algorithms derived in the previous chapters.

6.1 COST FUNCTION

Let us rewrite the cost function given in Chapter 5 in a slightly different form:

$$J(n) = D\left[\hat{\mathbf{h}}(n), \hat{\mathbf{h}}(n-1)\right] + \eta(n)\varepsilon^2(n), \tag{6.1}$$

where, again, $D\left[\hat{\mathbf{h}}(n), \hat{\mathbf{h}}(n-1)\right]$ is some measure of distance from the old to the new weight vectors and $\eta(n)$ is a positive variable parameter that usually depends on the input signal $x(n)$. The magnitude of $\eta(n)$ represents the importance of correctiveness compared to the importance of conservativeness (41). If $\eta(n)$ is very small, minimizing $J(n)$ is close to minimizing $D\left[\hat{\mathbf{h}}(n), \hat{\mathbf{h}}(n-1)\right]$, so that the algorithm makes very small updates. On the other hand, if $\eta(n)$ is very large, the minimization of $J(n)$ is almost equivalent to minimizing $D\left[\hat{\mathbf{h}}(n), \hat{\mathbf{h}}(n-1)\right]$, subject to the constraint $\varepsilon(n) = 0$.

To minimize $J(n)$, we need to set its L partial derivatives $\partial J(n)/\partial \hat{h}_l(n)$ to zero. Hence, the different weight coefficients $\hat{h}_l(n)$, $l = 0, 1, \ldots, L - 1$, will be found by solving the equations

$$\frac{\partial D\left[\hat{\mathbf{h}}(n), \hat{\mathbf{h}}(n-1)\right]}{\partial \hat{h}_l(n)} - 2\eta(n)x(n-l)\varepsilon(n) = 0. \tag{6.2}$$

For some distances (like the one used to derive the exponentiated gradient algorithm), (6.2) can be highly nonlinear so that solving it is very difficult if not impossible. However, if the new weight vector $\hat{\mathbf{h}}(n)$ is close to the old weight vector $\hat{\mathbf{h}}(n-1)$, replacing the a posteriori error signal $\varepsilon(n)$ in (6.2) with the a priori error signal $e(n)$ is a reasonable approximation and the equation

$$\frac{\partial D\left[\hat{\mathbf{h}}(n), \hat{\mathbf{h}}(n-1)\right]}{\partial \hat{h}_l(n)} - 2\eta(n)x(n-l)e(n) = 0 \tag{6.3}$$

is much easier to solve for all distance measures $D\left[\hat{\mathbf{h}}(n), \hat{\mathbf{h}}(n-1)\right]$.

The exponentiated gradient (EG) algorithms were first proposed by Kivinen and Warmuth in the context of computational learning theory (41). These algorithms are highly nonlinear and can be easily derived from the previous criterion, by simply using for the distance $D_{\text{re}}\left[\hat{\mathbf{h}}(n), \hat{\mathbf{h}}(n-1)\right]$, the relative entropy also known as Kullback-Leibler divergence. Since this divergence is not really a distance, it has to be handled with care.

6.2   THE EG ALGORITHM FOR POSITIVE WEIGHTS

In this section, we assume that the components of the impulse response that we need to identify are all positive, in order that the relative entropy is meaningful. Thus, we have

$$D_{\text{re}}\left[\hat{\mathbf{h}}(n), \hat{\mathbf{h}}(n-1)\right] = \sum_{l=0}^{L-1} \hat{h}_l(n) \ln \frac{\hat{h}_l(n)}{\hat{h}_l(n-1)}. \tag{6.4}$$

With this divergence measure, $\hat{\mathbf{h}}(n)$ and $\hat{\mathbf{h}}(n-1)$ are probability vectors, which means that their components are nonnegative and $\left\|\hat{\mathbf{h}}(n)\right\|_1 = \left\|\hat{\mathbf{h}}(n-1)\right\|_1 = \mathcal{H} > 0$, where $\mathcal{H}$ is a scaling factor. Therefore, we should minimize $J(n)$ with the constraint that $\sum_l \hat{h}_l(n) = 1$ (i.e., we take here $\mathcal{H} = 1$). This optimization leads to

$$\ln \frac{\hat{h}_l(n)}{\hat{h}_l(n-1)} + 1 - 2\eta(n)x(n-l)e(n) + \ell_{\text{m}} = 0, \; l = 0, 1, \ldots, L - 1, \tag{6.5}$$

where ℓ_{m} is a Lagrange multiplier. We then deduce the EG algorithm (41):

$$\hat{h}_l(n) = \frac{\hat{h}_l(n-1)r_l(n)}{\sum_{i=0}^{L-1} \hat{h}_i(n-1)r_i(n)}, \; l = 0, 1, \ldots, L - 1, \tag{6.6}$$

where

$$r_l(n) = \exp\left[2\eta(n)x(n-l)e(n)\right]. \tag{6.7}$$

The algorithm is initialized with $\hat{h}_l(0) = c > 0, \ \forall l$.

6.3   THE EG± ALGORITHM FOR POSITIVE AND NEGATIVE WEIGHTS

The EG algorithm is designed to work for positive weights only, due to the nature of the relative entropy definition. However, there is a simple way to generalize the idea to both positive and negative weights. Indeed, we can always find two vectors $\hat{\mathbf{h}}^+(n)$ and $\hat{\mathbf{h}}^-(n)$ with positive coefficients, in such a way that the vector

$$\hat{\mathbf{h}}(n) = \hat{\mathbf{h}}^+(n) - \hat{\mathbf{h}}^-(n) \tag{6.8}$$

can have positive and negative components. In this case, the a priori and a posteriori error signals can be written as

$$
\begin{aligned}
e(n) &= d(n) - \left[\hat{\mathbf{h}}^+(n-1) - \hat{\mathbf{h}}^-(n-1)\right]^T \mathbf{x}(n), & (6.9) \\
\varepsilon(n) &= d(n) - \left[\hat{\mathbf{h}}^+(n) - \hat{\mathbf{h}}^-(n)\right]^T \mathbf{x}(n), & (6.10)
\end{aligned}
$$

and the criterion (6.1) will change to (7), (8)

$$J^{\pm}(n) = D_{\text{re}}\left[\hat{\mathbf{h}}^+(n), \hat{\mathbf{h}}^+(n-1)\right] + D_{\text{re}}\left[\hat{\mathbf{h}}^-(n), \hat{\mathbf{h}}^-(n-1)\right] + \frac{\eta(n)}{\mathcal{H}}\varepsilon^2(n), \tag{6.11}$$

where $\mathcal{H}$ is a positive scaling constant. Using the Kullback-Leibler divergence plus the constraint $\sum_l[\hat{h}_l^+(k) + \hat{h}_l^-(k)] = \mathcal{H}$ and the approximation discussed in Section 6.1, the minimization of (6.11) gives

$$\ln\frac{\hat{h}_l^+(n)}{\hat{h}_l^+(n-1)} + 1 - \frac{2\eta(n)}{\mathcal{H}}x(n-l)e(n) + \ell_{\mathrm{m}} = 0, \tag{6.12}$$

$$\ln\frac{\hat{h}_l^-(n)}{\hat{h}_l^-(n-1)} + 1 + \frac{2\eta(n)}{\mathcal{H}}x(n-l)e(n) + \ell_{\mathrm{m}} = 0, \tag{6.13}$$
$$l = 0, 1, \ldots, L-1,$$

where ℓ_m is a Lagrange multiplier. From the two previous equations, we easily find the EG$\pm$ algorithm (41):

$$\hat{h}_l^+(n) = \mathcal{H}\frac{\hat{h}_l^+(n-1)r_l^+(n)}{\sum_{i=0}^{L-1}\left[\hat{h}_i^+(n-1)r_i^+(n) + \hat{h}_i^-(n-1)r_i^-(n)\right]}, \tag{6.14}$$

$$\hat{h}_l^-(n) = \mathcal{H}\frac{\hat{h}_l^-(n-1)r_l^-(n)}{\sum_{i=0}^{L-1}\left[\hat{h}_i^+(n-1)r_i^+(n) + \hat{h}_i^-(n-1)r_i^-(n)\right]}, \tag{6.15}$$

$$l = 0, 1, \ldots, L-1,$$

where

$$r_l^+(n) = \exp\left[\frac{2\eta(n)}{\mathcal{H}}x(n-l)e(n)\right], \tag{6.16}$$

$$r_l^-(n) = \exp\left[-\frac{2\eta(n)}{\mathcal{H}}x(n-l)e(n)\right] \tag{6.17}$$

$$= \frac{1}{r_l^+(n)}, \quad l = 0, 1, \ldots, L-1.$$

In the rest, we take $\eta(n) = L\alpha/\left[2\mathbf{x}^T(n)\mathbf{x}(n)\right]$. We can check that we always have $\left\|\hat{\mathbf{h}}^+(n)\right\|_1 + \left\|\hat{\mathbf{h}}^-(n)\right\|_1 = \mathcal{H}$. The fact that

$$\mathcal{H} = \left\|\hat{\mathbf{h}}^+(n)\right\|_1 + \left\|\hat{\mathbf{h}}^-(n)\right\|_1 \geq \left\|\hat{\mathbf{h}}^+(n) - \hat{\mathbf{h}}^-(n)\right\|_1 = \left\|\hat{\mathbf{h}}(n)\right\|_1 \tag{6.18}$$

suggests that the constant $\mathcal{H}$ has to be chosen such that $\mathcal{H} \geq \|\mathbf{h}\|_1$ in order that $\hat{\mathbf{h}}(n)$ converges to $\mathbf{h}$. If we take $\mathcal{H} < \|\mathbf{h}\|_1$, the algorithm will introduce a bias in the coefficients of the filter.

The EG$\pm$ algorithm is summarized in Table 6.1.

The motivation for the EG$\pm$ (or EG) algorithm can be developed by taking the logarithmic of (6.14) and (6.15). This shows that the logarithmic weights use almost the same update as the NLMS algorithm. Alternatively, this can be interpreted as exponentiating the update, hence the name EG$\pm$. This has the effect of assigning larger relative updates to larger weights, thereby deemphasizing the effect of smaller weights. This is qualitatively similar to the PNLMS algorithm which makes the update proportional to the size of the weight. This type of behavior is desirable for sparse impulse responses where small weights do not contribute significantly to the mean solution but introduce an undesirable noise-like variance.

In (34), it is shown that the excess MSE for the EG$\pm$ algorithm when $\mathcal{H} \approx \|\mathbf{h}\|_1$ is

$$J_{\text{ex}}(\infty) = \frac{\alpha}{2}\sigma_w^2\mathcal{H}\left(1 - \frac{\|\mathbf{h}\|_2^2}{\|\mathbf{h}\|_1^2}\right), \tag{6.19}$$

Table 6.1: The EG± algorithm.

Initialization:	$\hat{h}_l^+(0) = \hat{h}_l^-(0) = c > 0, \; l = 0, 1, \ldots, L - 1$
Parameters:	$\mathcal{H} \geq \|\mathbf{h}\|_1$
	$0 < \alpha < 2$
	$\delta_{\mathrm{EG}} = \mathrm{cst} \cdot \sigma_x^2$
Error:	$e(n) = d(n) - \left[\hat{\mathbf{h}}^+(n-1) - \hat{\mathbf{h}}^-(n-1) \right]^T \mathbf{x}(n)$
Update:	$\mu(n) = \dfrac{\alpha}{\mathbf{x}^T(n)\mathbf{x}(n) + \delta_{\mathrm{EG}}}$
	$r_l^+(n) = \exp\left[L \dfrac{\mu(n)}{\mathcal{H}} x(n-l)e(n) \right]$
	$r_l^-(n) = \dfrac{1}{r_l^+(n)}$
	$\hat{h}_l^+(n) = \mathcal{H} \dfrac{\hat{h}_l^+(n-1)r_l^+(n)}{\sum_{i=0}^{L-1}\left[\hat{h}_i^+(n-1)r_i^+(n) + \hat{h}_i^-(n-1)r_i^-(n) \right]}$
	$\hat{h}_l^-(n) = \mathcal{H} \dfrac{\hat{h}_l^-(n-1)r_l^-(n)}{\sum_{i=0}^{L-1}\left[\hat{h}_i^+(n-1)r_i^+(n) + \hat{h}_i^-(n-1)r_i^-(n) \right]}$
	$l = 0, 1, \ldots, L - 1$

where α is the normalized step-size parameter (see Table 6.1) and σ_w^2 is the variance of the noise. Expression (6.19) can be written as a function of the sparseness measure:

$$J_{\mathrm{ex}}(\infty) = \frac{\alpha}{2}\sigma_w^2 \mathcal{H} \left\{ 1 - \frac{1}{\left[\sqrt{L} - \left(\sqrt{L} - 1 \right) \xi_{12}(\mathbf{h}) \right]^2} \right\}. \tag{6.20}$$

We see from (6.20) that when the impulse response is very sparse, the term in the brackets is very small, and the normalized step size can be taken larger than the NLMS. As a result, the EG± can converge much faster than the NLMS with the same excess MSE.

6.4   LINK BETWEEN NLMS AND EG± ALGORITHMS

If we initialize $\hat{h}_l(0) = 0, \; l = 0, 1, \ldots, L - 1$, in the (non-regularized) NLMS algorithm, we can easily verify that

$$\begin{aligned} \hat{\mathbf{h}}(n) &= \sum_{j=0}^{n-1} \mu(j+1)\mathbf{x}(j+1)e(j+1) \\ &= \alpha \sum_{j=0}^{n-1} \frac{\mathbf{x}(j+1)e(j+1)}{\mathbf{x}^T(j+1)\mathbf{x}(j+1)}, \end{aligned} \tag{6.21}$$

where $\mu(j+1) = \alpha/\left[\mathbf{x}^T(j+1)\mathbf{x}(j+1)\right]$.

If we start the adaptation of the EG$\pm$ algorithm with $\hat{h}_l^+(0) = \hat{h}_l^-(0) = c > 0$, $l = 0, 1, \ldots, L-1$, we can show that (6.14) and (6.15) are equivalent to (7), (8)

$$\hat{h}_l^+(n) = \mathcal{H}\frac{t_l^+(n)}{\sum_{i=0}^{L-1}[t_i^+(n) + t_i^-(n)]}, \tag{6.22}$$

$$\hat{h}_l^-(n) = \mathcal{H}\frac{t_l^-(n)}{\sum_{i=0}^{L-1}[t_i^+(n) + t_i^-(n)]}, \tag{6.23}$$

where

$$t_l^+(n) = \exp\left[\frac{2}{\mathcal{H}}\sum_{j=0}^{n-1}\eta(j+1)x(j+1-l)e(j+1)\right], \tag{6.24}$$

$$t_l^-(n) = \exp\left[-\frac{2}{\mathcal{H}}\sum_{i=0}^{n-1}\eta(j+1)x(j+1-l)e(j+1)\right] \tag{6.25}$$

$$= \frac{1}{t_l^+(n)},$$

and $\eta(j+1) = L\alpha/\left[2\mathbf{x}^T(j+1)\mathbf{x}(j+1)\right]$. Clearly, the convergence of the algorithm does not depend on the initialization parameter c (as long as it is positive and nonzero). Now

$$\hat{h}_l(n) = \hat{h}_l^+(n) - \hat{h}_l^-(n)$$

$$= \mathcal{H}\frac{t_l^+(n) - t_l^-(n)}{\sum_{i=0}^{L-1}[t_i^+(n) + t_i^-(n)]}$$

$$= \mathcal{H}\frac{\sinh\left[\frac{2}{\mathcal{H}}\sum_{j=0}^{n-1}\eta(j+1)x(j+1-l)e(j+1)\right]}{\sum_{i=0}^{L-1}\cosh\left[\frac{2}{\mathcal{H}}\sum_{j=0}^{n-1}\eta(j+1)x(j+1-i)e(j+1)\right]}. \tag{6.26}$$

Note that the sinh function has the effect of exponentiating the update, as previously commented.

For $\mathcal{H}$ large enough and using the approximations $\sinh(a) \approx a$ and $\cosh(a) \approx 1$ when $|a| \ll 1$, (6.26) becomes

$$\hat{h}_l(n) = \alpha\sum_{j=0}^{n-1}\frac{x(j+1-l)e(j+1)}{\mathbf{x}^T(j+1)\mathbf{x}(j+1)}, \quad 0 \le l \le L-1. \tag{6.27}$$

Comparing (6.21) and (6.27), we understand that for $\mathcal{H}$ large enough, the NLMS and EG$\pm$ algorithms have the same performance. Obviously, the choice of $\mathcal{H}$ is critical in practice: if we take

$\mathcal{H} < \|\mathbf{h}\|_1$, the EG± will introduce a bias in the coefficients of the filter and if $\mathcal{H} \gg \|\mathbf{h}\|_1$, the EG± will behave like NLMS.

6.5   LINK BETWEEN IPNLMS AND EG± ALGORITHMS

PNLMS and IPNLMS algorithms were developed for use in network echo cancelers (5). In comparison to the NLMS algorithm, they have very fast initial convergence and tracking when the echo path is sparse. As previously mentioned, the idea behind these "proportionate" algorithms is to update each coefficient of the filter independently of the others by adjusting the adaptation step size in proportion to the estimated filter coefficient.

How are the IPNLMS and EG± algorithms specifically related? In the rest of this section, we show that the IPNLMS is in fact an approximation of the EG±.

If we suppose that $\hat{\mathbf{h}}^{+}(n)$ [resp. $\hat{\mathbf{h}}^{-}(n)$] is close to $\hat{\mathbf{h}}^{+}(n-1)$ [resp. $\hat{\mathbf{h}}^{-}(n-1)$], which is usually the case in all adaptive algorithms (especially for a small step size), the two distances $D_{\text{re}}\left[\hat{\mathbf{h}}^{+}(n), \hat{\mathbf{h}}^{+}(n-1)\right]$ and $D_{\text{re}}\left[\hat{\mathbf{h}}^{-}(n), \hat{\mathbf{h}}^{-}(n-1)\right]$ in criterion (6.11) can be approximated as follows:

$$
\begin{aligned}
D_{\text{re}}\left[\hat{\mathbf{h}}^{+}(n), \hat{\mathbf{h}}^{+}(n-1)\right] &= \sum_{l=0}^{L-1} \hat{h}_l^{+}(n) \ln \frac{\hat{h}_l^{+}(n)}{\hat{h}_l^{+}(n-1)} \\
&\approx \sum_{l=0}^{L-1} \hat{h}_l^{+}(n) \left[\frac{\hat{h}_l^{+}(n)}{\hat{h}_l^{+}(n-1)} - 1\right],
\end{aligned}
\tag{6.28}
$$

$$
\begin{aligned}
D_{\text{re}}[\hat{\mathbf{h}}^{-}(n), \hat{\mathbf{h}}^{-}(n-1)] &= \sum_{l=0}^{L-1} \hat{h}_l^{-}(n) \ln \frac{\hat{h}_l^{-}(n)}{\hat{h}_l^{-}(n-1)} \\
&\approx \sum_{l=0}^{L-1} \hat{h}_l^{-}(n) \left[\frac{\hat{h}_l^{-}(n)}{\hat{h}_l^{-}(n-1)} - 1\right].
\end{aligned}
\tag{6.29}
$$

Using (6.28) and (6.29) plus the constraint $\sum_l \left[\hat{h}_l^{+}(k) + \hat{h}_l^{-}(k)\right] = \mathcal{H}$ and the same approximation as for the EG±, the minimization of (6.11) gives the approximated EG± algorithm:

$$
\hat{h}_l^{+}(n) = \hat{h}_l^{+}(n-1)\left[1 + \frac{\eta(n)}{\mathcal{H}}x(n-l)e(n) - \frac{\eta(n)}{\mathcal{H}^2}\hat{y}(n)e(n)\right],
\tag{6.30}
$$

$$
\hat{h}_l^{-}(n) = \hat{h}_l^{-}(n-1)\left[1 - \frac{\eta(n)}{\mathcal{H}}x(n-l)e(n) - \frac{\eta(n)}{\mathcal{H}^2}\hat{y}(n)e(n)\right],
\tag{6.31}
$$

so that

$$
\begin{aligned}
\hat{h}_l(n) &= \hat{h}_l^+(n) - \hat{h}_l^-(n) \\
&= \hat{h}_l(n-1) + \frac{\eta(n)\left[\hat{h}_l^+(n-1) + \hat{h}_l^-(n-1)\right]}{\mathcal{H}} x(n-l)e(n) \\
&\quad - \frac{\eta(n)}{\mathcal{H}^2}\hat{h}_l(n-1)\hat{y}(n)e(n).
\end{aligned}
\tag{6.32}
$$

Neglecting the last term in the right-hand side of (6.32), we get

$$
\hat{h}_l(n) = \hat{h}_l(n-1) + \eta(n) \cdot \frac{\hat{h}_l^+(n-1) + \hat{h}_l^-(n-1)}{\left\|\hat{\mathbf{h}}^+(n-1)\right\|_1 + \left\|\hat{\mathbf{h}}^-(n-1)\right\|_1} x(n-l)e(n).
\tag{6.33}
$$

If the true impulse response $\mathbf{h}$ is sparse, it can be shown that if we choose $\mathcal{H} = \|\mathbf{h}\|_1$, the (positive) vector $\hat{\mathbf{h}}^+(n-1) + \hat{\mathbf{h}}^-(n-1)$ is also sparse after convergence. This means that the elements

$$
\frac{\hat{h}_l^+(n-1) + \hat{h}_l^-(n-1)}{\left\|\hat{\mathbf{h}}^+(n-1)\right\|_1 + \left\|\hat{\mathbf{h}}^-(n-1)\right\|_1}
$$

in (6.33) play exactly the same role as the elements $g_l(n-1)$ in the IPNLMS algorithm, in the particular case where $\kappa = 1$ (PNLMS algorithm). As a result, we can expect the two algorithms (IPNLMS and EG$\pm$) to have similar performance. On the other hand, if $\mathcal{H} \gg \|\mathbf{h}\|_1$, it can be shown that $\hat{h}_l^+(n-1) + \hat{h}_l^-(n-1) \approx \mathcal{H}/L$, $\forall l$. In this case, the EG$\pm$ algorithm will behave like IPNLMS with $\kappa = -1$ (NLMS algorithm). Thus, the parameter κ in IPNLMS operates like the parameter $\mathcal{H}$ in EG$\pm$. However, the advantage of IPNLMS is that no a priori information of the system impulse response is required in order to have a better convergence rate than the NLMS algorithm. Another clear advantage of IPNLMS is that it is much less complex to implement than EG$\pm$. We conclude that IPNLMS is a good approximation of EG$\pm$ and is more useful in practice. Note also that the approximated EG$\pm$ algorithm (6.33) belongs to the family of natural gradient algorithms (26), (49).

CHAPTER 7

The Mu-Law PNLMS and Other PNLMS-Type Algorithms

Many interesting adaptive algorithms have been developed for the identification of sparse impulse responses. Sparse adaptive filters were mainly designed in the context of network echo cancellation. The idea of exploiting the sparseness character of the echo paths has appeared in the nineties, e.g., (35), (48), (69). However, the PNLMS algorithm proposed by Duttweiler in 2000 (18) can be considered as a milestone in the field. The main feature of this algorithm is that it does not require any a priori information about the echo path, since the direction of update is based only on the current filter estimate. This is a very important issue in real-world/real-time applications such as network and acoustic echo cancellation. Inspired by the "proportionate" idea, many PNLMS-type algorithms were proposed in the last decade.

The basic PNLMS-type algorithms have been reviewed in Chapter 5. The goal of this chapter is to present some of the latest developments in the field of sparse adaptive filters.

7.1   THE MU-LAW PNLMS ALGORITHMS

The idea behind the PNLMS algorithm is to update each coefficient of the filter independently of the others by adjusting the adaptation step size in proportion to the magnitude of the estimated filter coefficient. The adaptation gains are redistributed among all the coefficients, emphasizing the large ones in order to speed up their convergence, thus achieving a fast initial convergence rate. Unfortunately, after this initial phase, the convergence rate of the PNLMS algorithm slows down significantly, even becoming slower than NLMS. This is due to the fact that the equations used to calculate the step-size control factors (i.e., the proportionate factors) are not based on any optimization criteria but are designed in an ad-hoc way. In order to improve the performance of the PNLMS algorithm, the natural approach is to try finding an optimal criterion to evaluate these proportionate factors.

A very interesting idea was developed in (14), (15), by following the steepest-descent method (33) adapted for PNLMS-type algorithms. Considering an adaptive filter of length L, defined by

$\hat{\mathbf{h}}(n)$, the steepest-descent algorithm with the step-size control matrix can be written as (15)

$$\hat{\mathbf{h}}(n) = \hat{\mathbf{h}}(n-1) + \mu \mathbf{G}(n-1) \left[\mathbf{p}_{\mathbf{x}d} - \mathbf{R}_{\mathbf{x}} \hat{\mathbf{h}}(n-1) \right], \tag{7.1}$$

where μ is the step-size parameter, $\mathbf{G}(n-1)$ is the diagonal step-size control matrix (containing the proportionate factors), $\mathbf{p}_{\mathbf{x}d} = E[\mathbf{x}(n)d(n)]$ is the cross-correlation vector between the input vector and the desired signal, and $\mathbf{R}_{\mathbf{x}} = E[\mathbf{x}^T(n)\mathbf{x}(n)]$ is the correlation matrix of the input vector. The Wiener-Hopf equations (33) state that

$$\mathbf{R}_{\mathbf{x}}\mathbf{p}_{\mathbf{x}d} = \hat{\mathbf{h}}_{\mathrm{W}}, \tag{7.2}$$

where $\hat{\mathbf{h}}_{\mathrm{W}}$ denotes the optimal Wiener filter (see Section 4.1 in Chapter 4). Using (7.2) in (7.1), we get

$$\hat{\mathbf{h}}(n) = \hat{\mathbf{h}}(n-1) + \mu \mathbf{G}(n-1)\mathbf{R}_{\mathbf{x}} \left[\hat{\mathbf{h}}_{\mathrm{W}} - \hat{\mathbf{h}}(n-1) \right]. \tag{7.3}$$

Defining the misalignment vector as

$$\mathbf{m}(n) = \hat{\mathbf{h}}_{\mathrm{W}} - \hat{\mathbf{h}}(n), \tag{7.4}$$

the update (7.3) can be rewritten as

$$\mathbf{m}(n) = \left[\mathbf{I} - \mu \mathbf{G}(n-1)\mathbf{R}_{\mathbf{x}} \right] \mathbf{m}(n-1). \tag{7.5}$$

In general, the natural way to initialize the filter coefficients is $\hat{\mathbf{h}}(0) = \mathbf{0}$, so that $\mathbf{m}(0) = \hat{\mathbf{h}}_{\mathrm{W}}$. Consequently, by expanding (7.5) results

$$\mathbf{m}(n) = \prod_{k=1}^{n} \left[\mathbf{I} - \mu \mathbf{G}(k-1)\mathbf{R}_{\mathbf{x}} \right] \hat{\mathbf{h}}_{\mathrm{W}}. \tag{7.6}$$

In order to further facilitate the analysis, it is assumed that the input signal is a white Gaussian noise, so that $\mathbf{R}_{\mathbf{x}} = \sigma_x^2 \mathbf{I}$, where σ_x^2 is the input signal variance. In this case, (7.6) becomes

$$\mathbf{m}(n) = \prod_{k=1}^{n} \left[\mathbf{I} - \psi \mathbf{G}(k-1) \right] \hat{\mathbf{h}}_{\mathrm{W}}, \tag{7.7}$$

where $\psi = \mu \sigma_x^2$. Therefore, the elements of the misalignment vector are given by

$$m_l(n) = \prod_{k=1}^{n} [1 - \psi g_l(k-1)] \hat{h}_{\mathrm{W},l}, \ l = 0, 1, \ldots, L-1, \tag{7.8}$$

where $g_l(n-1)$ is the $(l+1)$th diagonal element of the matrix $\mathbf{G}(n-1)$, i.e., the step-size control factor associated with the coefficient $\hat{h}_l(n-1)$ of the adaptive filter. In the following, the condition $\sum_{l=0}^{L-1} g_l(n-1) = L$ is assumed.

It can be considered that each coefficient of the adaptive filter reaches the ζ-vicinity of its optimal value (for a given small positive number ζ) in a specific number of iterations, i.e.,

$$\prod_{k=1}^{n_l} [1 - \psi g_l(k-1)] \hat{h}_{\mathrm{W},l} = \zeta, \quad l = 0, 1, \ldots, L-1. \tag{7.9}$$

It is clear that $n_\mathrm{T} = \max(n_0, n_1, \ldots, n_{L-1})$ iterations are needed for all the coefficients to reach the ζ-vicinity of their optimal values. The main problem is how to calculate the proportionate factors in order to achieve the fastest overall convergence, i.e., to make n_T minimum. It was demonstrated in (15) that under the condition $0 < \psi g_l(n-1) \ll 1$, for any n, the minimum number of iterations for convergence is

$$n_{\mathrm{T,min}} = \frac{1}{\psi L} \sum_{l=0}^{L-1} \ln \frac{|\hat{h}_{\mathrm{W},l}|}{\zeta}. \tag{7.10}$$

On the other hand, assuming that fixed proportionate factors are used, i.e., $g_l(n-1) = g_l$, for any n, the number of iterations n_l from (7.9) can be determined as

$$n_l = \frac{\ln \dfrac{|\hat{h}_{\mathrm{W},l}|}{\zeta}}{\ln \dfrac{1}{1 - \psi g_l}}, \quad l = 0, 1, \ldots, L-1. \tag{7.11}$$

Also, under the condition $0 < \psi g_l \ll 1$, it can be shown that $\ln(1 - \psi g_l) \approx -\psi g_l$. Thus, (7.11) can be rewritten as

$$n_l \approx \frac{\ln \dfrac{|\hat{h}_{\mathrm{W},l}|}{\zeta}}{\psi g_l}, \quad l = 0, 1, \ldots, L-1. \tag{7.12}$$

Ideally, $n_l = n_{\mathrm{T,min}}$, for $l = 0, 1, \ldots, L-1$, i.e., all the coefficients reach the ζ-vicinity of their optimal values after the same number of iterations. Under this condition, taking (7.10) and (7.12) into account, the proportionate factors result as

$$g_l \approx \frac{\ln \dfrac{|\hat{h}_{\mathrm{W},l}|}{\zeta}}{\dfrac{1}{L} \sum_{i=0}^{L-1} \ln \dfrac{|\hat{h}_{\mathrm{W},i}|}{\zeta}}, \quad l = 0, 1, \ldots, L-1. \tag{7.13}$$

It was also shown in (15), that (7.13) can be further approximated as

$$
\begin{aligned}
g_l &\approx \frac{\ln\left(1 + \frac{|\hat{h}_{W,l}|}{\zeta}\right)}{\frac{1}{L}\sum_{i=0}^{L-1}\ln\left(1 + \frac{|\hat{h}_{W,i}|}{\zeta}\right)} \\
&= \frac{F\left(|\hat{h}_{W,l}|\right)}{\frac{1}{L}\sum_{i=0}^{L-1} F\left(|\hat{h}_{W,i}|\right)}, \quad l = 0, 1, \ldots, L-1.
\end{aligned}
\tag{7.14}
$$

It can be noticed that the function $F(\cdot)$ is in fact the mu-law used for non-uniform compression in telecommunication applications (13). This particular aspect gives the name of the algorithm developed based on (7.14), i.e., the mu-law PNLMS (MPNLMS) (14), (15). Of course, in practice, $\hat{h}_l(n-1)$ is used instead of $\hat{h}_{W,l}$, with $l = 0, 1, \ldots, L-1$. The MPNLMS algorithm is obtained following the equations of the PNLMS algorithm (see Table 5.1 in Chapter 5) and is summarized in Table 7.1.

It is known that the logarithm function is an expensive operation to implement in most platforms. The authors in (15) proposed to use a piecewise linear function instead of the logarithmic one, i.e.,

$$
F(a) = \begin{cases} 600a, & \text{if } a < 0.005 \\ \\ 3, & \text{otherwise} \end{cases}.
\tag{7.15}
$$

The corresponding algorithm is called the segment PNLMS (SPNLMS) (15), and it can be considered as a more practical version of the MPNLMS algorithm.

The parameter ζ used in the MPNLMS algorithm is a very small positive number. Its value should be chosen based on the system noise level (i.e., the noise that corrupts the output of the echo path). In echo cancellation, a recommended value could be $\zeta = 0.001$ because the echo below -60 dB is negligible (15). However, the overall performance of the MPNLMS algorithm can be improved by using a variable parameter, $\zeta(n)$ (71). The idea is to start the algorithm using a large value for $\zeta(n)$ and slowly decrease the required ζ-vicinity to be reached by the converged algorithm. The value of $\zeta(n)$ is determined based on the current estimate of the mean square error, i.e.,

$$
\hat{\sigma}_e^2(n) = \lambda\hat{\sigma}_e^2(n-1) + (1-\lambda)e^2(n),
\tag{7.16}
$$

where $0 < \lambda \leq 1$ and $e(n)$ is the error signal. Then, the variable parameter is evaluated as

$$
\zeta(n) = \sqrt{\frac{\varphi\hat{\sigma}_e^2(n)}{L\sigma_x^2}},
\tag{7.17}
$$

Table 7.1: The mu-law PNLMS (MPNLMS) algorithm.

Initialization:	$\hat{h}_l(0) = 0, \ l = 0, 1, \ldots, L - 1$
Parameters:	$\zeta = 0.001$
	$\delta_{\mathrm{p}} = 0.01$
	$\varrho = 5/L$
	$0 < \alpha < 2$
	$\delta_{\mathrm{MPNLMS}} = \mathrm{cst} \cdot \sigma_x^2$
Error:	$e(n) = d(n) - \mathbf{x}^T(n)\hat{\mathbf{h}}(n - 1)$

Update:

$$F\left[\left|\hat{h}_l(n - 1)\right|\right] = \ln\left[1 + \frac{\left|\hat{h}_l(n - 1)\right|}{\zeta}\right], \ l = 0, 1, \ldots, L - 1$$

$$\gamma_{\min}(n - 1) = \varrho \max\left\{\delta_{\mathrm{p}}, \ F\left[\left|\hat{h}_0(n - 1)\right|\right], \ \ldots, \ F\left[\left|\hat{h}_{L-1}(n - 1)\right|\right]\right\}$$

$$\gamma_l(n - 1) = \max\left\{\gamma_{\min}(n - 1), \ F\left[\left|\hat{h}_l(n - 1)\right|\right]\right\}$$

$$g_l(n - 1) = \frac{\gamma_l(n - 1)}{\frac{1}{L}\sum_{i=0}^{L-1}\gamma_i(n - 1)}, \ l = 0, 1, \ldots, L - 1$$

$$\mu(n) = \frac{\alpha}{\sum_{i=0}^{L-1} x^2(n - i)g_i(n - 1) + \delta_{\mathrm{MPNLMS}}}$$

$$\hat{h}_l(n) = \hat{h}_l(n - 1) + \mu(n)g_l(n - 1)x(n - l)e(n),$$
$$l = 0, 1, \ldots, L - 1$$

where φ is a small positive number (and the recommended value is $\varphi = 0.001$). The resulting algorithm is called the adaptive MPNLMS (AMPNLMS) (71).

7.2   THE SPARSENESS-CONTROLLED PNLMS ALGORITHMS

The MPNLMS algorithm discussed in the previous section uses the optimal step-size control factors to achieve, under some conditions, the fastest overall convergence until the adaptive filter reaches its steady state. This algorithm was derived such that all coefficients attain a converged value within the ζ-vicinity of their optimal value in the same number of iterations.

However, similar to the PNLMS algorithm, the MPNLMS suffers from slow convergence when the echo path is dispersive (e.g., like in acoustic echo cancellation). In order to improve the performance of these algorithms when dealing with such situations, a novel approach was recently proposed in (45). Let us recall the sparseness measure based on the l_1 and l_2 norms [see (2.26) in

Section 2.4, Chapter 2], i.e.,

$$\xi_{12}(\mathbf{h}) = \frac{L}{L - \sqrt{L}} \left(1 - \frac{\|\mathbf{h}\|_1}{\sqrt{L}\|\mathbf{h}\|_2} \right), \tag{7.18}$$

where $\mathbf{h}$ denotes the true impulse response that needs to be identified. Since $\mathbf{h}$ is unavailable in practice, an estimated sparseness measure can be evaluated using the coefficients of the adaptive filter, i.e.,

$$\xi_{12}\left[\hat{\mathbf{h}}(n) \right] = \frac{L}{L - \sqrt{L}} \left[1 - \frac{\|\hat{\mathbf{h}}(n)\|_1}{\sqrt{L}\|\hat{\mathbf{h}}(n)\|_2} \right]. \tag{7.19}$$

The further goal is to incorporate the measure from (7.19) into PNLMS-type algorithms, in order to improve their robustness to varying levels of sparseness of impulse responses. Let us remember that the PNLMS algorithm uses a constant ϱ (see Table 5.1 in Chapter 5) that prevents the coefficients from stalling when they are much smaller than the largest coefficient; the typical value is $\varrho = 5/L$. However, it is reasonable to use a high value of ϱ when estimating dispersive impulse responses, i.e., when $\xi_{12}[\hat{\mathbf{h}}(n)]$ is small. The approach proposed in (45) considers a variable parameter $\varrho(n)$, as a function of the sparseness measure from (7.19), evaluated as

$$\varrho(n) = \begin{cases} 5/L, & \text{if } n < L \\ \\ e^{-\rho\xi_{12}[\hat{\mathbf{h}}(n)]}, & \text{if } n \geq L \end{cases}, \tag{7.20}$$

thus achieving the previous discussed effect. The constant ρ in (7.20) influences the overall performance. A large value of ρ is suitable when we need to identify very sparse impulse responses. On the other hand, the value of ρ should be small when identifying dispersive impulse responses. It was shown in (45) that a good compromise choice is $\rho = 6$. The variable parameter from (7.20) can be incorporated directly in both the PNLMS and MPNLMS algorithms, thus resulting the sparseness-controlled PNLMS (SC-PNLMS) and the sparseness-controlled MPNLMS (SC-MPNLMS) algorithms (45). For example, the SC-PNLMS algorithm is summarized in Table 7.2. It was also shown in (45) that the sparseness-controlled idea can be adapted to other PNLMS-type algorithms.

7.3   THE PNLMS ALGORITHM WITH INDIVIDUAL ACTIVATION FACTORS

The overall performance of the regular PNLMS algorithm (18) depends on some predefined parameters controlling initialization and proportionality (see Table 5.1 in Chapter 5), i.e., 1) the constant δ_{p} (typically $\delta_{\mathrm{p}} = 0.01$), which regularizes the update when all coefficients are zero at initialization and 2) the parameter ϱ (usually $\varrho = 5/L$) that prevents the coefficients from stalling when they are much smaller than the magnitude of the largest coefficient. These parameters influence the overall convergence behavior of the PNLMS algorithm (18), so that it is important to set them

---

Table 7.2: The sparseness-controlled PNLMS (SC-PNLMS) algorithm.

Initialization:	$\hat{h}_l(0) = 0,\ l = 0, 1, \ldots, L - 1$
Parameters:	$\rho = 6$
	$\delta_{\mathrm{p}} = 0.01$
	$\varrho(n) = 5/L,\ n < L$
	$0 < \alpha < 2$
	$\delta_{\mathrm{SC-PNLMS}} = \mathrm{cst} \cdot \sigma_x^2/L$
Error:	$e(n) = d(n) - \mathbf{x}^T(n)\hat{\mathbf{h}}(n-1)$
Update:	$\gamma_{\min}(n-1) = \varrho(n-1) \max\left[\delta_{\mathrm{p}},\ \left|\hat{h}_0(n-1)\right|,\ \ldots,\ \left|\hat{h}_{L-1}(n-1)\right|\right]$
	$\gamma_l(n-1) = \max\left[\gamma_{\min}(n-1),\ \left|\hat{h}_l(n-1)\right|\right]$
	$g_l(n-1) = \dfrac{\gamma_l(n-1)}{\sum_{i=0}^{L-1} \gamma_i(n-1)},\ l = 0, 1, \ldots, L - 1$
	$\mu(n) = \dfrac{\alpha}{\sum_{i=0}^{L-1} x^2(n-i)g_i(n-1) + \delta_{\mathrm{SC-PNLMS}}}$
	$\hat{h}_l(n) = \hat{h}_l(n-1) + \mu(n)g_l(n-1)x(n-l)e(n),$
	$\qquad l = 0, 1, \ldots, L - 1$
	$\xi_{12}\left[\hat{\mathbf{h}}(n)\right] = \dfrac{L}{L - \sqrt{L}}\left[1 - \dfrac{\|\hat{\mathbf{h}}(n)\|_1}{\sqrt{L}\|\hat{\mathbf{h}}(n)\|_2}\right]$
	$\varrho(n) = e^{-\rho \xi_{12}[\hat{\mathbf{h}}(n)]},\ n \geq L$

properly. However, constant values for these parameters could limit somehow the performance of the algorithm. Let us remember that the sparseness-controlled algorithms (45) described in the previous section use a variable parameter $\varrho(n)$, as a function of an estimated sparseness measure, thus improving the performance of PNLMS-type algorithms when estimating dispersive impulse responses. Very recently, another approach was proposed in (68), following the idea of using variable parameters in a more general manner.

Let us recall the basic formulas of the PNLMS algorithm, which are used to compute the proportionate factors (see Chapter 5), i.e.,

$$\gamma_{\min}(n-1) = \varrho \max\left[\delta_{\mathrm{p}},\ \left|\hat{h}_0(n-1)\right|,\ \ldots,\ \left|\hat{h}_{L-1}(n-1)\right|\right], \tag{7.21}$$

$$\gamma_l(n-1) = \max\left[\gamma_{\min}(n-1),\ \left|\hat{h}_l(n-1)\right|\right], \tag{7.22}$$

$$g_l(n-1) = \frac{\gamma_l(n-1)}{\sum_{i=0}^{L-1} \gamma_i(n-1)},\ l = 0, 1, \ldots, L - 1. \tag{7.23}$$

The authors in (68) referred to the parameter from the left-hand side of (7.21) as the "activation factor." According to (7.21), it can be noticed that the activation factor is common to all coefficients and depends on $\left\|\hat{\mathbf{h}}(n-1)\right\|_\infty$ (see Chapter 2). In this way, according to (7.22), two situations can be identified:

1) if $\gamma_{\min}(n-1) > \left|\hat{h}_l(n-1)\right|$, then the coefficient $\hat{h}_l(n-1)$ is *inactive* and its associated gain (i.e., proportionate factor) is

$$g_l(n-1) \;=\; \frac{\gamma_{\min}(n-1)}{\sum_{i=0}^{L-1}\gamma_i(n-1)}, \qquad (7.24)$$

2) if $\gamma_{\min}(n-1) \le \left|\hat{h}_l(n-1)\right|$, then the coefficient $\hat{h}_l(n-1)$ is *active* and its associated gain is

$$g_l(n-1) \;=\; \frac{\left|\hat{h}_l(n-1)\right|}{\sum_{i=0}^{L-1}\gamma_i(n-1)}. \qquad (7.25)$$

In the first situation, the activation factor is common to all filter coefficients, so that a minimum and common gain given in (7.24) is assigned to all inactive coefficients. It was demonstrated in (68) that this is an undesired feature of the regular PNLMS algorithm because the gain from (7.24) is not proportional with its associated coefficient $\hat{h}_l(n-1)$. On the other hand, in the second situation, each gain evaluated as in (7.25) is associated with the magnitude of the active coefficient, being proportional to $\left|\hat{h}_l(n-1)\right|$.

In accordance with the proportionate concept, the goal is to make the gain proportional to its associated coefficient, even when this one is inactive. Consequently, each gain assigned to an inactive coefficient will be assigned with an individual activation factor instead of a common one. Therefore, (7.22) becomes

$$\gamma_l(n-1) = \max\left[\gamma_{\min,l}(n-1),\ \left|\hat{h}_l(n-1)\right|\right]. \qquad (7.26)$$

The solution proposed in (68) is to compute the individual activation factors as

$$\gamma_{\min,l}(n) = \chi\left|\hat{h}_l(n)\right| + (1-\chi)\gamma_l(n-1), \qquad (7.27)$$

where $0 < \chi < 1$ and $\gamma_{\min,l}(0) = 0.01/L$. The parameter χ plays the role of a forgetting factor related to the memory of the adaptive filter coefficient magnitude. In practice, the recommended choice is $\chi = 1/2$ (since there is no a priori information about the system to be identified). Also, because the estimated coefficient $\hat{h}_l(n)$ may not be proportional to its corresponding true value (i.e., h_l) at the beginning of the adaptation process, the solution is to periodically update the individual activation factors $\gamma_{\min,l}(n)$ only after a learning period of L samples (i.e., the adaptive filter length). Taking the previous considerations into account, (7.27) is rewritten as

$$\gamma_{\min,l}(n) = \begin{cases} \dfrac{1}{2}\left|\hat{h}_l(n)\right| + \dfrac{1}{2}\gamma_l(n-1), & \text{if } n = L, 2L, 3L, \ldots \\[2mm] \gamma_{\min,l}(n-1), & \text{otherwise} \end{cases} \qquad (7.28)$$

Table 7.3: The individual activation factor PNLMS (IAF-PNLMS) algorithm.

Initialization:	$\hat{h}_l(0) = 0, \ l = 0, 1, \ldots, L-1$
	$\gamma_{\min,l}(0) = 0.01/L, \ l = 0, 1, \ldots, L-1$
Parameters:	$0 < \alpha < 2$
	$\delta_{\text{IAF-PNLMS}} = \text{cst} \cdot \sigma_x^2/L$
Error:	$e(n) = d(n) - \mathbf{x}^T(n)\hat{\mathbf{h}}(n-1)$
Update:	$\gamma_l(n-1) = \max\left[\gamma_{\min,l}(n-1), \ \left|\hat{h}_l(n-1)\right|\right]$

$$g_l(n-1) = \frac{\gamma_l(n-1)}{\sum_{i=0}^{L-1} \gamma_i(n-1)}, \ l = 0, 1, \ldots, L-1$$

$$\mu(n) = \frac{\alpha}{\sum_{i=0}^{L-1} x^2(n-i)g_i(n-1) + \delta_{\text{IAF-PNLMS}}}$$

$$\hat{h}_l(n) = \hat{h}_l(n-1) + \mu(n)g_l(n-1)x(n-l)e(n),$$
$$l = 0, 1, \ldots, L-1$$

$$\gamma_{\min,l}(n) = \begin{cases} \dfrac{1}{2}\left|\hat{h}_l(n)\right| + \dfrac{1}{2}\gamma_l(n-1), & \text{if } n = L, 2L, 3L, \ldots \\[2ex] \gamma_{\min,l}(n-1), & \text{otherwise} \end{cases}$$

The resulted algorithm is named the individual activation factor PNLMS (IAF-PNLMS) (68) and is summarized in Table 7.3.

CHAPTER 8

Variable Step-Size PNLMS Algorithms

The overall performance of PNLMS-type algorithms is controlled by the step-size parameter. It is known that a constant value of the normalized step-size parameter leads to a compromise between fast convergence and good tracking ability on the one hand, and low misadjustment on the other hand. This is the basic feature inherited from the NLMS algorithm. In order to meet these conflicting requirements, the normalized step size needs to be controlled. This was the motivation behind the development of variable step-size NLMS (VSS-NLMS) algorithms. Consequently, it is expected that VSS techniques can also improve the performance of PNLMS-type algorithms.

In the first part of this chapter, we briefly outline the influence of the normalized step-size parameter over the performance of NLMS-based algorithms. The second part of this chapter is dedicated to the development of a simple and practical VSS-PNLMS algorithm, following the approach presented in Chapter 4, Section 4.4.

8.1 CONSIDERATIONS ON THE CONVERGENCE OF THE NLMS ALGORITHM

NLMS-based algorithms are widely used in practice due to their simplicity and numerical robustness. Basically, there are two parameters that need to be tuned within the NLMS algorithm, i.e., the normalized step-size parameter and the regularization constant. According to their values, we can control the performance of the algorithm in terms of convergence rate, tracking ability, and misadjustment. In this section, we provide a brief convergence analysis in order to outline the influence of these parameters over the performance of the NLMS algorithm and to motivate the needs for using VSS techniques.

Let us consider the classical problem of system identification (33). The reference signal is defined as

$$d(n) = \mathbf{h}^T \mathbf{x}(n) + w(n), \tag{8.1}$$

where $\mathbf{h}$ denotes the impulse response vector of the unknown system [a finite-impulse-response (FIR) filter of length L], $\mathbf{x}(n)$ is a vector containing the most recent L samples of the input signal, and $w(n)$ is the system noise (assumed to be white in this section). The well-known update of the

NLMS algorithm is

$$\hat{\mathbf{h}}(n) = \hat{\mathbf{h}}(n-1) + \frac{\alpha \mathbf{x}(n) e(n)}{\delta + \mathbf{x}^T(n) \mathbf{x}(n)}, \tag{8.2}$$

where α is the normalized step-size parameter, δ is the regularization constant, and

$$e(n) = d(n) - \hat{\mathbf{h}}^T(n-1) \mathbf{x}(n) \tag{8.3}$$

is the error signal. The misalignment vector is defined as the difference between the true coefficients of the system and the adaptive filter coefficients, i.e.,

$$\mathbf{m}(n) = \mathbf{h} - \hat{\mathbf{h}}(n). \tag{8.4}$$

Consequently, (8.2) becomes

$$\mathbf{m}(n) = \mathbf{m}(n-1) - \frac{\alpha \mathbf{x}(n) e(n)}{\delta + \mathbf{x}^T(n) \mathbf{x}(n)}. \tag{8.5}$$

Taking the square of the ℓ_2 norm in (8.5), we obtain

$$\|\mathbf{m}(n)\|_2^2 = \|\mathbf{m}(n-1)\|_2^2 - \frac{2\mathbf{x}^T(n) \mathbf{m}(n-1) e(n)}{\delta + \mathbf{x}^T(n) \mathbf{x}(n)} \alpha + \frac{\mathbf{x}^T(n) \mathbf{x}(n) e^2(n)}{[\delta + \mathbf{x}^T(n) \mathbf{x}(n)]^2} \alpha^2. \tag{8.6}$$

Next, taking the expectation in (8.6), we have

$$\begin{aligned}
E\left[\|\mathbf{m}(n)\|_2^2\right] &= E\left[\|\mathbf{m}(n-1)\|_2^2\right] - 2E\left[\frac{\mathbf{x}^T(n) \mathbf{m}(n-1) e(n)}{\delta + \mathbf{x}^T(n) \mathbf{x}(n)}\right] \alpha \\
&\quad + E\left\{\frac{\mathbf{x}^T(n) \mathbf{x}(n) e^2(n)}{[\delta + \mathbf{x}^T(n) \mathbf{x}(n)]^2}\right\} \alpha^2.
\end{aligned} \tag{8.7}$$

In order to facilitate the analysis, let us assume that

$$\mathbf{x}^T(n) \mathbf{x}(n) = \|\mathbf{x}(n)\|_2^2 \approx L\sigma_x^2, \tag{8.8}$$

which is true for $L \gg 1$ and where σ_x^2 denotes the input signal power. Therefore, $\mathbf{x}^T(n) \mathbf{x}(n)$ is considered as a constant. Using (8.1), (8.3), and (8.4), the numerator of the second term from the right-hand side of (8.7) can be expressed as

$$\begin{aligned}
&E\left[\mathbf{x}^T(n) \mathbf{m}(n-1) e(n)\right] \\
&= E\left[\mathbf{x}^T(n) \mathbf{m}(n-1) w(n) + \mathbf{m}^T(n-1) \mathbf{x}(n) \mathbf{x}^T(n) \mathbf{m}(n-1)\right].
\end{aligned} \tag{8.9}$$

Since the system noise is uncorrelated with the input signal and is assumed to be white, (8.9) becomes

$$\begin{aligned}
E\left[\mathbf{x}^T(n) \mathbf{m}(n-1) e(n)\right] &= E\left[\mathbf{m}^T(n-1) \mathbf{x}(n) \mathbf{x}^T(n) \mathbf{m}(n-1)\right] \\
&= E\left\{\mathrm{tr}\left[\mathbf{m}(n-1) \mathbf{m}^T(n-1) \mathbf{x}(n) \mathbf{x}^T(n)\right]\right\}.
\end{aligned} \tag{8.10}$$

In the following, we assume that the input signal is a white Gaussian noise, so that

$$E\left[\mathbf{x}\left(n\right)\mathbf{x}^{T}\left(n\right)\right] = \sigma_{x}^{2}\mathbf{I}. \tag{8.11}$$

Consequently, with the independence assumption (33), (8.10) can be rewritten as

$$
\begin{aligned}
E\left[\mathbf{x}^{T}\left(n\right)\mathbf{m}\left(n-1\right)e\left(n\right)\right] &= \mathrm{tr}\left\{E\left[\mathbf{m}\left(n-1\right)\mathbf{m}^{T}\left(n-1\right)\right]E\left[\mathbf{x}\left(n\right)\mathbf{x}^{T}\left(n\right)\right]\right\} \\
&= \sigma_{x}^{2}E\left[\left\|\mathbf{m}\left(n-1\right)\right\|_{2}^{2}\right],
\end{aligned}
\tag{8.12}
$$

so that [based on (8.8) and (8.12)] the second term from the right-hand side of (8.7) can be approximated by

$$E\left[\frac{\mathbf{x}^{T}\left(n\right)\mathbf{m}\left(n-1\right)e\left(n\right)}{\delta+\mathbf{x}^{T}\left(n\right)\mathbf{x}\left(n\right)}\right] \approx \frac{\sigma_{x}^{2}}{\delta+L\sigma_{x}^{2}}E\left[\left\|\mathbf{m}\left(n-1\right)\right\|_{2}^{2}\right]. \tag{8.13}$$

Similarly, the last term of (8.7) becomes

$$E\left\{\frac{\mathbf{x}^{T}\left(n\right)\mathbf{x}\left(n\right)e^{2}\left(n\right)}{\left[\delta+\mathbf{x}^{T}\left(n\right)\mathbf{x}\left(n\right)\right]^{2}}\right\} \approx L\sigma_{x}^{2}\frac{E\left[e^{2}\left(n\right)\right]}{\left(\delta+L\sigma_{x}^{2}\right)^{2}}. \tag{8.14}$$

In the same way, we can evaluate

$$
\begin{aligned}
E\left[e^{2}\left(n\right)\right] &= E\left\{\left[w\left(n\right)+\mathbf{m}^{T}\left(n-1\right)\mathbf{x}\left(n\right)\right]^{2}\right\} \\
&= \sigma_{w}^{2} + E\left[\mathbf{m}^{T}\left(n-1\right)\mathbf{x}\left(n\right)\mathbf{x}^{T}\left(n\right)\mathbf{m}\left(n-1\right)\right] \\
&\approx \sigma_{w}^{2} + \sigma_{x}^{2}E\left[\left\|\mathbf{m}\left(n-1\right)\right\|_{2}^{2}\right],
\end{aligned}
\tag{8.15}
$$

where σ_{w}^{2} is the power of the system noise. Based on (8.13) and (8.15), (8.7) becomes

$$
\begin{aligned}
E\left[\left\|\mathbf{m}\left(n\right)\right\|_{2}^{2}\right] &\approx E\left[\left\|\mathbf{m}\left(n-1\right)\right\|_{2}^{2}\right] - 2\frac{\sigma_{x}^{2}E\left[\left\|\mathbf{m}\left(n-1\right)\right\|_{2}^{2}\right]}{\delta+L\sigma_{x}^{2}}\alpha \\
&\quad + L\sigma_{x}^{2}\frac{\sigma_{w}^{2}+\sigma_{x}^{2}E\left[\left\|\mathbf{m}\left(n-1\right)\right\|_{2}^{2}\right]}{\left(\delta+L\sigma_{x}^{2}\right)^{2}}\alpha^{2}.
\end{aligned}
\tag{8.16}
$$

Thus,

$$
\begin{aligned}
E\left[\left\|\mathbf{m}\left(n\right)\right\|_{2}^{2}\right] &\approx \left[1 - \frac{2\sigma_{x}^{2}}{\delta+L\sigma_{x}^{2}}\alpha + \frac{L\sigma_{x}^{4}}{\left(\delta+L\sigma_{x}^{2}\right)^{2}}\alpha^{2}\right]E\left[\left\|\mathbf{m}\left(n-1\right)\right\|_{2}^{2}\right] \\
&\quad + \frac{L\sigma_{x}^{2}\sigma_{w}^{2}}{\left(\delta+L\sigma_{x}^{2}\right)^{2}}\alpha^{2}.
\end{aligned}
\tag{8.17}
$$

Let us denote

$$f\left(\alpha, \delta, L, \sigma_x^2\right) = 1 - \frac{2\sigma_x^2}{\delta + L\sigma_x^2}\alpha + \frac{L\sigma_x^4}{\left(\delta + L\sigma_x^2\right)^2}\alpha^2, \tag{8.18}$$

$$g\left(\alpha, \delta, L, \sigma_x^2, \sigma_w^2\right) = \frac{L\sigma_x^2\sigma_w^2}{\left(\delta + L\sigma_x^2\right)^2}\alpha^2, \tag{8.19}$$

so that (8.17) can be resumed as

$$E\left[\|\mathbf{m}\left(n\right)\|_2^2\right] \approx f\left(\alpha, \delta, L, \sigma_x^2\right) E\left[\|\mathbf{m}\left(n-1\right)\|_2^2\right] + g\left(\alpha, \delta, L, \sigma_x^2, \sigma_w^2\right). \tag{8.20}$$

The result from (8.20) illustrates a "separation" between the convergence and the misadjustment components. Therefore, the term $f\left(\alpha, \delta, L, \sigma_x^2\right)$ influences the convergence rate of the algorithm. As expected, it depends on the normalized step-size value, the regularization constant, the filter length, and the input signal power. It is interesting to notice that it does not depend on the system noise power. Besides, some classical conclusions can be established by analyzing the behavior of the convergence term. First, it can be noticed that the fastest convergence (FC) mode is obtained when the function from (8.18) reaches its minimum. Taking the normalized step size as the reference parameter, we obtain

$$\alpha_{\text{FC}} = 1 + \frac{\delta}{L\sigma_x^2}. \tag{8.21}$$

Neglecting the regularization constant (i.e., $\delta = 0$), the fastest convergence mode is achieved for $\alpha = 1$, which is a well-known result. Second, the stability condition can be found by imposing $|f(\alpha, \delta, L, \sigma_x^2)| < 1$, which leads to

$$0 < \alpha_{\text{stable}} < 2\left(1 + \frac{\delta}{L\sigma_x^2}\right). \tag{8.22}$$

Also, taking $\delta = 0$ in (8.22), the classical stability condition of the NLMS algorithm results, i.e., $0 < \alpha < 2$.

The second term, $g(\alpha, \delta, L, \sigma_x^2, \sigma_w^2)$, influences the misadjustment of the algorithm and it depends on the system noise power. Hence, the convergence rate of the algorithm is not influenced by the level of the system noise, but the misadjustment increases when the system noise increases. More importantly, it can be noticed that the misadjustment term from (8.19) always increases when α increases; this concludes the fact that a higher value of the normalized step size increases the misadjustment. From this point of view, in order to achieve the lowest misadjustment (LM), we need to take

$$\alpha_{\text{LM}} \approx 0. \tag{8.23}$$

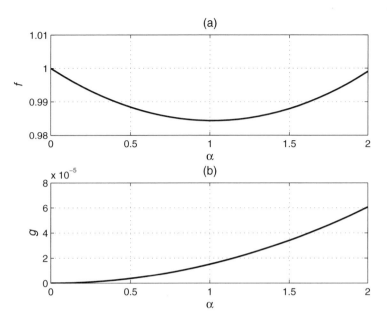

Figure 8.1: (a) Evolution of the convergence term f from (8.18), as a function of the normalized step-size parameter α. (b) Evolution of the misadjustment term g from (8.19), as a function of the normalized step-size parameter α. The regularization constant is $\delta = \sigma_x^2$.

Nevertheless, the ideal requirements of the algorithm are for both fast convergence and low misadjustment. It is clear that the conditions (8.21) and (8.23) "push" the normalized step size in opposite directions. This aspect is the motivation for the VSS approaches, i.e., the normalized step size needs to be controlled in order to meet these conflicting requirements.

The regularization constant also influences the performance of the algorithm, but in a "milder" way. It can be noticed that the convergence term from (8.18) always decreases when the regularization constant increases, while the misadjustment term from (8.19) always increases when the regularization constant decreases.

The evolution of the terms from (8.18) and (8.19), as a function of the normalized step-size parameter and the regularization constant are depicted in Figs. 8.1 and 8.2, respectively. The parameters were set to $\sigma_x^2 = 1, \sigma_w^2 = 0.001$, and $L = 64$. These plots support the previously discussed theoretical aspects.

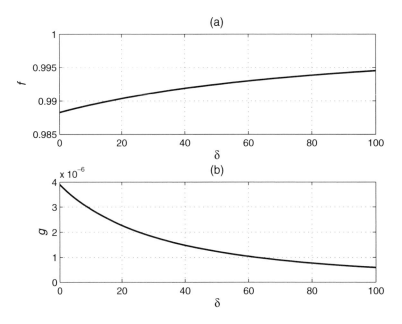

Figure 8.2: (a) Evolution of the convergence term f from (8.18), as a function of the regularization constant δ. (b) Evolution of the misadjustment term g from (8.19), as a function of the regularization constant δ. The normalized step-size parameter is $\alpha = 0.5$.

8.2   A VARIABLE STEP-SIZE PNLMS ALGORITHM

The general update of PNLMS-type algorithms is

$$\hat{\mathbf{h}}(n) = \hat{\mathbf{h}}(n-1) + \frac{\alpha \mathbf{G}(n-1)\mathbf{x}(n)e(n)}{\delta + \mathbf{x}^T(n)\mathbf{G}(n-1)\mathbf{x}(n)}, \tag{8.24}$$

where

$$\mathbf{G}(n-1) = \mathrm{diag}\left[\; g_0(n-1) \quad g_1(n-1) \quad \cdots \quad g_{L-1}(n-1) \;\right] \tag{8.25}$$

is an $L \times L$ diagonal matrix [see (5.9)–(5.11) in Section 5.1, Chapter 5]. Using this matrix, an individual step size is assigned to each filter coefficient, in such a way that a larger coefficient receives a larger increment, thus increasing the convergence rate of that coefficient. Let us define the weighted input vector

$$\begin{aligned}
\mathbf{x}_g(n) &= \left[\; g_0^{1/2}(n-1)x(n) \quad g_1^{1/2}(n-1)x(n-1) \quad \cdots \quad g_{L-1}^{1/2}(n-1)x(n-L+1) \;\right]^T \\
&= \left[\; x_g(n) \quad x_g(n-1) \quad \cdots \quad x_g(n-L+1) \;\right]^T.
\end{aligned} \tag{8.26}$$

Consequently, the update (8.24) becomes

$$\hat{\mathbf{h}}(n) = \hat{\mathbf{h}}(n-1) + \frac{\alpha \mathbf{G}^{1/2}(n-1)\mathbf{x}_g(n)e(n)}{\delta + \mathbf{x}_g^T(n)\mathbf{x}_g(n)}. \tag{8.27}$$

Looking at (8.27), it is interesting to notice that a PNLMS-type algorithm can be also interpreted as a sort of a variable step-size technique, since a time-variant normalized step size is used (but individual for each filter coefficient).

In order to develop a VSS-PNLMS algorithm, we will follow the approach from Section 4.4, Chapter 4, which provides a simple and elegant VSS technique. Let us rewrite the update (8.24) as

$$\hat{\mathbf{h}}(n) = \hat{\mathbf{h}}(n-1) + \mu(n)\mathbf{G}(n-1)\mathbf{x}(n)e(n). \tag{8.28}$$

Considering the same system identification scenario from the previous section, our goal is to find an expression for $\mu(n)$ such that $E[\varepsilon^2(n)] = \sigma_w^2$ [see (4.85)], where

$$\varepsilon(n) = d(n) - \hat{\mathbf{h}}^T(n)\mathbf{x}(n) \tag{8.29}$$

is the a posteriori error signal. In this manner, we aim to recover the system noise from the error of the adaptive filter, which is a reasonable approach in system identification problems. Using (8.28) in (8.29) and taking (8.3) and (8.26) into account, we find that

$$\begin{aligned}
\varepsilon(n) &= d(n) - \left[\hat{\mathbf{h}}^T(n-1) + \mu(n)\mathbf{x}^T(n)\mathbf{G}(n-1)e(n)\right]\mathbf{x}(n) \\
&= \left[d(n) - \hat{\mathbf{h}}^T(n-1)\mathbf{x}(n)\right] - \mu(n)\mathbf{x}^T(n)\mathbf{G}(n-1)\mathbf{x}(n)e(n) \\
&= \left[1 - \mu(n)\mathbf{x}^T(n)\mathbf{G}(n-1)\mathbf{x}(n)\right]e(n) \\
&= \left[1 - \mu(n)\mathbf{x}_g^T(n)\mathbf{x}_g(n)\right]e(n).
\end{aligned} \tag{8.30}$$

Squaring and taking the expectation in (8.30), assuming that the input and error signals are uncorrelated (which is true when the adaptive filter has started to converge to the true solution), and using the approximation $\mathbf{x}_g^T(n)\mathbf{x}_g(n) = LE\left[x_g^2(n)\right] = L\sigma_{x_g}^2$ for $L \gg 1$, we get

$$E[\varepsilon^2(n)] = \left[1 - 2\mu(n)L\sigma_{x_g}^2 + \mu^2(n)\left(L\sigma_{x_g}^2\right)^2\right]\sigma_e^2(n), \tag{8.31}$$

where $E[e^2(n)] = \sigma_e^2(n)$ is the variance of the error signal. Finally, imposing the condition $E[\varepsilon^2(n)] = \sigma_w^2$ in (8.31), we obtain the quadratic equation

$$\mu^2(n) - \frac{2}{L\sigma_{x_g}^2}\mu(n) + \frac{1}{\left(L\sigma_{x_g}^2\right)^2}\left[1 - \frac{\sigma_w^2}{\sigma_e^2(n)}\right] = 0, \tag{8.32}$$

from which the obvious solution is

$$\mu(n) = \frac{1}{\mathbf{x}_g^T(n)\mathbf{x}_g(n)}\left[1 - \frac{\sigma_w}{\sigma_e(n)}\right].$$

$$(8.33)$$

For practical reasons, the step size from (8.33) needs to be evaluated as

$$\mu(n) = \begin{cases} \dfrac{1}{\delta + \mathbf{x}^T(n)\mathbf{G}(n-1)\mathbf{x}(n)}\left[1 - \dfrac{\sigma_w}{\epsilon + \hat{\sigma}_e(n)}\right], & \text{if } \hat{\sigma}_e(n) \geq \sigma_w \\ 0, & \text{otherwise} \end{cases},$$

$$(8.34)$$

where δ is the regularization constant, ϵ is a very small positive number to avoid division by zero, and the variance of the error signal is estimated as

$$\hat{\sigma}_e^2(n) = \lambda\hat{\sigma}_e^2(n-1) + (1-\lambda)e^2(n),$$

$$(8.35)$$

where λ is an exponential window [its value is chosen as $\lambda = 1 - 1/(KL)$, with $K \geq 2$] and the initial value is $\hat{\sigma}_e^2(0) = 0$. The reason for using the second line in (8.34) is the following. Theoretically, it is clear that $\sigma_e(n) \geq \sigma_w$, which implies that $\mu(n) \geq 0$. Nevertheless, the estimation from (8.35) could result in a lower magnitude than σ_w^2, which would make $\mu(n)$ negative. Thus, in this situation, the problem is solved by setting $\mu(n) = 0$.

Using the step size from (8.34) in (8.28), we obtain a variable step-size PNLMS-type algorithm. The only a priori parameter needed by this algorithm is the power of the system noise, σ_w^2; in echo cancellation context it can be easily estimated during silences. Looking at (8.34) it is obvious that before the algorithm converges, $\hat{\sigma}_e(n)$ is large compared to σ_w and, consequently, the normalized step size is close to 1, which provides the fastest convergence. When the algorithm starts to converge to the true solution, $\hat{\sigma}_e(n) \approx \sigma_w$ and $\mu(n) \approx 0$. In fact, this is the desired behavior for the adaptive algorithm, leading to both good convergence and low misadjustment.

CHAPTER 9

Proportionate Affine Projection Algorithms

The affine projection algorithm (APA) can be interpreted as a generalization of the NLMS algorithm. The main advantage of the APA over the NLMS algorithm consists of a superior convergence rate, especially for correlated inputs (like speech). For this reason, the APA and different versions of it were found to be very attractive choices for echo cancellation. Since PNLMS-type algorithms usually outperform the NLMS algorithm for sparse impulse responses, it was found natural to combine the "proportionate" ideas with the APA, thus resulting the proportionate APAs (PAPAs).

Most of PAPAs were derived based on a straightforward extension of PNLMS-type algorithms, as explained in the first part of this chapter. In the second section, we present a more rigorous way to derive PAPAs, which will lead to an improved PAPA, in terms of both computational complexity and convergence performance. Finally, the last part of this chapter is dedicated to a variable step-size (VSS) technique that further increases the performance of these algorithms.

9.1 CLASSICAL DERIVATION

The APA [originally proposed in (54)] was derived as a generalization of the NLMS algorithm, in the sense that each tap weight vector update of the NLMS is viewed as a one dimensional affine projection, while in the APA the projections are made in multiple dimensions. When the projection dimension increases, the convergence rate of the tap weight vector also increases. However, this also leads to an increased computational complexity. Considering an FIR adaptive filter of length L, defined by the coefficients vector $\hat{\mathbf{h}}(n)$, the equations that define the classical APA are (54)

$$\mathbf{e}(n) = \mathbf{d}(n) - \mathbf{X}^T(n)\hat{\mathbf{h}}(n-1), \tag{9.1}$$

$$\hat{\mathbf{h}}(n) = \hat{\mathbf{h}}(n-1) + \alpha\mathbf{X}(n)\left[\delta\mathbf{I}_P + \mathbf{X}^T(n)\mathbf{X}(n)\right]^{-1}\mathbf{e}(n), \tag{9.2}$$

where

$$\mathbf{d}(n) = \left[\begin{array}{cccc} d(n) & d(n-1) & \cdots & d(n-P+1) \end{array}\right]^T$$

is a vector containing the most recent P samples of the reference signal, with P denoting the projection order, the matrix

$$\mathbf{X}(n) = \left[\begin{array}{cccc} \mathbf{x}(n) & \mathbf{x}(n-1) & \cdots & \mathbf{x}(n-P+1) \end{array}\right]$$

is the input signal matrix, with

$$\mathbf{x}(n-p) = \begin{bmatrix} x(n-p) & x(n-p-1) & \cdots & x(n-p-L+1) \end{bmatrix}^T, \quad p = 0, 1, \ldots, P-1$$

being the input signal vectors, the constant α denotes the step-size parameter, δ is the regularization constant, and $\mathbf{I}_P$ is the $P \times P$ identity matrix. It can be easily noticed that for $P = 1$ the NLMS algorithm is obtained.

Let us recall now the update of PNLMS-type algorithms, i.e.,

$$\hat{\mathbf{h}}(n) = \hat{\mathbf{h}}(n-1) + \frac{\alpha \mathbf{G}(n-1)\mathbf{x}(n)e(n)}{\delta + \mathbf{x}^T(n)\mathbf{G}(n-1)\mathbf{x}(n)}, \tag{9.3}$$

where

$$e(n) = d(n) - \hat{\mathbf{h}}^T(n-1)\mathbf{x}(n) \tag{9.4}$$

is the error signal and

$$\mathbf{G}(n-1) = \mathrm{diag}\begin{bmatrix} g_0(n-1) & g_1(n-1) & \cdots & g_{L-1}(n-1) \end{bmatrix} \tag{9.5}$$

is an $L \times L$ diagonal matrix [see (5.9)–(5.11) in Section 5.1, Chapter 5] that assigns an individual step size to each filter coefficient (thus, a larger coefficient receives a larger increment, which further increases the convergence rate of that coefficient). Looking at (9.2) and (9.3), it was found natural to combine these approaches in a very straightforward manner, thus resulting the basic update for most of PAPAs [e.g., (23), (72)], i.e.,

$$\hat{\mathbf{h}}(n) = \hat{\mathbf{h}}(n-1) + \alpha \mathbf{G}(n-1)\mathbf{X}(n)\left[\delta \mathbf{I}_P + \mathbf{X}^T(n)\mathbf{G}(n-1)\mathbf{X}(n)\right]^{-1}\mathbf{e}(n). \tag{9.6}$$

In practice, it would be very computationally expensive (and also inefficient) to compute the matrix product $\mathbf{G}(n-1)\mathbf{X}(n)$ in the classical way (i.e., matrices multiplication). Hence, taking into account the diagonal character of the matrix $\mathbf{G}(n-1)$, we can evaluate

$$\begin{aligned} \mathbf{P}(n) &= \mathbf{G}(n-1)\mathbf{X}(n) \\ &= \begin{bmatrix} \mathbf{g}(n-1)\odot\mathbf{x}(n) & \mathbf{g}(n-1)\odot\mathbf{x}(n-1) & \cdots & \mathbf{g}(n-1)\odot\mathbf{x}(n-P+1) \end{bmatrix}, \end{aligned} \tag{9.7}$$

where

$$\mathbf{g}(n-1) = \begin{bmatrix} g_0(n-1) & g_1(n-1) & \cdots & g_{L-1}(n-1) \end{bmatrix}^T \tag{9.8}$$

is a vector containing the diagonal elements of $\mathbf{G}(n-1)$ and the operator $\odot$ denotes the Hadamard product, i.e., $\mathbf{a}\odot\mathbf{b} = \begin{bmatrix} a_0 b_0 & a_1 b_1 & \cdots & a_{L-1}b_{L-1} \end{bmatrix}^T$, with $\mathbf{a}$ and $\mathbf{b}$ being two vectors of length L. Using (9.7), the PAPAs update (9.6) can be rewritten as

$$\hat{\mathbf{h}}(n) = \hat{\mathbf{h}}(n-1) + \alpha \mathbf{P}(n)\left[\delta \mathbf{I}_P + \mathbf{X}^T(n)\mathbf{P}(n)\right]^{-1}\mathbf{e}(n). \tag{9.9}$$

Clearly, when implementing this type of algorithm in practice, the second line of (9.7) is used, requiring PL multiplications for evaluating the matrix $\mathbf{P}(n)$. Thus, the equations that define the "classical" PAPAs are (9.1), (9.7), and (9.9).

9.2   A NOVEL DERIVATION

Several PAPAs were proposed, (23), (36), (44), (72), based on (9.6), which were straightforwardly obtained by a simple combination between the APA update (9.2) and the "proportionate" ideas (9.3). In the first part of this section, a more rigorous approach for deriving PAPAs is presented.

It is known that the classical APA can be derived by minimizing the squared Euclidian norm of the difference vector $\hat{\mathbf{h}}(n) - \hat{\mathbf{h}}(n-1)$, subject to a set of P constraints, i.e.,

$$\mathbf{d}(n) = \mathbf{X}^T(n)\hat{\mathbf{h}}(n). \tag{9.10}$$

In other words, the requirement is to cancel P a posteriori errors. In the following, it is assumed that the length of the adaptive filter is larger than the projection order, i.e., $L > P$. Using the method of Lagrange multipliers with multiple constraints, the cost function to derive the APA is

$$J(n) = \left\| \hat{\mathbf{h}}(n) - \hat{\mathbf{h}}(n-1) \right\|_2^2 + \left[\mathbf{d}(n) - \mathbf{X}^T(n)\hat{\mathbf{h}}(n) \right]^T \boldsymbol{\theta}(n), \tag{9.11}$$

where $\boldsymbol{\theta}(n) = \begin{bmatrix} \theta_0(n) & \theta_1(n) & \cdots & \theta_{P-1}(n) \end{bmatrix}^T$ is the Lagrange vector. Taking the gradient of $J(n)$ with respect to $\hat{\mathbf{h}}(n)$ and equating the result to zero, i.e., $\nabla_{\hat{\mathbf{h}}(n)} J(n) = \mathbf{0}$, we find that

$$\hat{\mathbf{h}}(n) = \hat{\mathbf{h}}(n-1) + \frac{1}{2}\mathbf{X}(n)\boldsymbol{\theta}(n). \tag{9.12}$$

On the other hand, using (9.10) in (9.1), we obtain

$$\mathbf{e}(n) = \mathbf{X}^T(n)\left[\hat{\mathbf{h}}(n) - \hat{\mathbf{h}}(n-1) \right]. \tag{9.13}$$

Then, introducing (9.12) in (9.13), the Lagrange vector is obtained as

$$\boldsymbol{\theta}(n) = 2\left[\mathbf{X}^T(n)\mathbf{X}(n) \right]^{-1}\mathbf{e}(n) \tag{9.14}$$

and, consequently, (9.12) becomes

$$\hat{\mathbf{h}}(n) = \hat{\mathbf{h}}(n-1) + \mathbf{X}(n)\left[\mathbf{X}^T(n)\mathbf{X}(n) \right]^{-1}\mathbf{e}(n). \tag{9.15}$$

For practical reasons, the step-size parameter α and the regularization matrix $\delta\mathbf{I}_P$ need to be introduced in (9.15), resulting the final update of the APA given in (9.2).

Looking again at (9.12), it can be noticed that each coefficient of the filter is updated as

$$\hat{h}_l(n) = \hat{h}_l(n-1) + \frac{1}{2}\mathbf{x}_P^T(n-l)\boldsymbol{\theta}(n), \ l = 0, 1, \ldots, L-1, \tag{9.16}$$

where

$$\mathbf{x}_P(n-l) = \begin{bmatrix} x(n-l) & x(n-l-1) & \cdots & x(n-l-P+1) \end{bmatrix}^T.$$

In order to obtain a PAPA, (9.16) should be rewritten as

$$\hat{h}_l(n) = \hat{h}_l(n-1) + \frac{1}{2}\mathbf{x}_P^T(n-l)\,\mathbf{G}_l(n-1)\,\boldsymbol{\theta}'(n),\tag{9.17}$$

where $\boldsymbol{\theta}'(n)$ is the new Lagrange vector. The $P \times P$ diagonal matrix $\mathbf{G}_l(n-1)$ is related to the coefficient $\hat{h}_l(n-1)$, with $l = 0, 1, \ldots, L-1$; this matrix "adjusts" (i.e., proportionates) the increment of the coefficient value. Taking (9.17) into account, (9.12) becomes

$$\hat{\mathbf{h}}(n) = \hat{\mathbf{h}}(n-1) + \frac{1}{2}\mathbf{X}'(n)\,\boldsymbol{\theta}'(n),\tag{9.18}$$

where the $L \times P$ matrix $\mathbf{X}'(n)$ is defined as

$$\mathbf{X}'(n) = \begin{bmatrix} \mathbf{x}_P^T(n)\,\mathbf{G}_0(n-1) \\ \mathbf{x}_P^T(n-1)\,\mathbf{G}_1(n-1) \\ \vdots \\ \mathbf{x}_P^T(n-L+1)\,\mathbf{G}_{L-1}(n-1) \end{bmatrix}.\tag{9.19}$$

Next, using (9.18) in (9.13), the new solution for $\boldsymbol{\theta}'(n)$ is

$$\boldsymbol{\theta}'(n) = 2\left[\mathbf{X}^T(n)\,\mathbf{X}'(n)\right]^{-1}\mathbf{e}(n).\tag{9.20}$$

Finally, the general update of the PAPA becomes

$$\hat{\mathbf{h}}(n) = \hat{\mathbf{h}}(n-1) + \alpha\mathbf{X}'(n)\left[\delta\mathbf{I}_P + \mathbf{X}^T(n)\,\mathbf{X}'(n)\right]^{-1}\mathbf{e}(n).\tag{9.21}$$

The main problem is how to choose the matrix $\mathbf{X}'(n)$, in terms of the diagonal matrices $\mathbf{G}_l(n-1)$, with $l = 0, 1, \ldots, L-1$. It is clear that when $\mathbf{G}_l(n-1) = \mathbf{I}_P$, the classical APA is obtained. Choosing $\mathbf{G}_l(n-1) = g_l(n-1)\mathbf{I}_P$, where $g_l(n-1)$ is the lth diagonal element of $\mathbf{G}(n-1)$ [see (9.5)], the matrix from the right-hand side of (9.19) is identical to $\mathbf{P}(n)$ defined in (9.7), i.e.,

$$\begin{bmatrix} g_0(n-1)x(n) & \cdots & g_0(n-1)x(n-P+1) \\ \vdots & \ddots & \vdots \\ g_{L-1}(n-1)x(n-L+1) & \cdots & g_{L-1}(n-1)x(n-P-L+2) \end{bmatrix}$$
$$= \mathbf{G}(n-1)\mathbf{X}(n)$$
$$= \begin{bmatrix} \mathbf{g}(n-1)\odot\mathbf{x}(n) & \mathbf{g}(n-1)\odot\mathbf{x}(n-1) & \cdots & \mathbf{g}(n-1)\odot\mathbf{x}(n-P+1) \end{bmatrix}$$
$$= \mathbf{P}(n)\tag{9.22}$$

and thus, the update (9.21) is the classical PAPA update from (9.9).

However, the APA can be viewed as an algorithm with "memory," i.e., it takes into account the "history" of the last P time samples. The classical PAPA, which is derived based on (9.22), does not take into account the "proportionate history" of each coefficient $\hat{h}_l(n-1)$, with $l = 0, 1, \ldots, L-1$, but only its proportionate factor from the current time sample, i.e., $g_l(n-1)$. Therefore, let us consider a modified approach in order to take advantage of the "proportionate memory" of the algorithm, by choosing the matrix (58)

$$\mathbf{G}_l(n-1) = \mathrm{diag}\begin{bmatrix} g_l(n-1) & g_l(n-2) & \cdots & g_l(n-P) \end{bmatrix}. \tag{9.23}$$

In this manner, we take into account the "proportionate history" of the coefficient $\hat{h}_l(n-1)$, in terms of its proportionate factors from the last P time samples. Thus, the matrix from the right-hand side of (9.19) becomes

$$\begin{bmatrix} g_0(n-1)x(n) & \cdots & g_0(n-P)x(n-P+1) \\ \vdots & \ddots & \vdots \\ g_{L-1}(n-1)x(n-L+1) & \cdots & g_{L-1}(n-P)x(n-P-L+2) \end{bmatrix} \tag{9.24}$$
$$= \begin{bmatrix} \mathbf{g}(n-1) \odot \mathbf{x}(n) & \mathbf{g}(n-2) \odot \mathbf{x}(n-1) & \cdots & \mathbf{g}(n-P) \odot \mathbf{x}(n-P+1) \end{bmatrix}$$
$$= \mathbf{P}'(n)$$

and, consequently, the update (9.21) is

$$\hat{\mathbf{h}}(n) = \hat{\mathbf{h}}(n-1) + \alpha \mathbf{P}'(n) \left[\delta \mathbf{I}_P + \mathbf{X}^T(n)\mathbf{P}'(n) \right]^{-1} \mathbf{e}(n). \tag{9.25}$$

We will refer to this algorithm as the "memory" PAPA (MPAPA).

The advantage of this modification is twofold. First, the MPAPA takes into account the "history" of the proportionate factors from the last P steps. The gain in terms of fast convergence and tracking could become more apparent when the projection order P increases. Second, the computational complexity is lower as compared with the classical PAPA. This is because the second line of (9.24) can be realized recursively as

$$\mathbf{P}'(n) = \begin{bmatrix} \mathbf{g}(n-1) \odot \mathbf{x}(n) & \mathbf{P}'_{-1}(n-1) \end{bmatrix}, \tag{9.26}$$

where the matrix

$$\mathbf{P}'_{-1}(n-1) = \begin{bmatrix} \mathbf{g}(n-2) \odot \mathbf{x}(n-1) & \cdots & \mathbf{g}(n-P) \odot \mathbf{x}(n-P+1) \end{bmatrix}$$

contains the first $P-1$ columns of $\mathbf{P}'(n-1)$. Thus, the columns from 1 to $P-1$ of the matrix $\mathbf{P}'(n-1)$ can be used directly for computing the matrix $\mathbf{P}'(n)$ [i.e., they become the columns from 2 to P of $\mathbf{P}'(n)$]. This is not the case of the classical PAPAs where all the columns of $\mathbf{P}(n)$ [see the second line of (9.7)] have to be evaluated at each iteration, because all of them are multiplied with the same vector $\mathbf{g}(n-1)$. Concluding, the evaluation of $\mathbf{P}(n)$ from (9.7) needs PL multiplications while the evaluation of $\mathbf{P}'(n)$ [see (9.26)] requires only L multiplications. This advantage becomes more apparent when the projection order P increases. Also, the fact that $\mathbf{P}'(n)$ has the time-shift property [like the data matrix $\mathbf{X}(n)$] could be a possible opportunity to establish a link with the fast APA (24), (70). It is also likely possible to derive efficient ways to compute the linear system involved in (9.25).

Besides, let us examine the matrix to be inverted in the classical PAPA, as compared to the case of the MPAPA. In the first case, taking the second line of (9.7) into account, the matrix to be inverted in (9.9) can be expressed as

$$
\begin{aligned}
\mathbf{M}(n) &= \delta \mathbf{I}_P + \mathbf{X}^T(n)\mathbf{P}(n) \\
&= \delta \mathbf{I}_P + \begin{bmatrix} \mathbf{x}^T(n) \\ \vdots \\ \mathbf{x}^T(n-P+1) \end{bmatrix} \begin{bmatrix} \mathbf{g}(n-1)\odot\mathbf{x}(n) & \cdots & \mathbf{g}(n-1)\odot\mathbf{x}(n-P+1) \end{bmatrix} \\
&= \begin{bmatrix} \delta + \mathbf{x}^T(n)[\mathbf{g}(n-1)\odot\mathbf{x}(n)] & \cdots & \mathbf{x}^T(n)[\mathbf{g}(n-1)\odot\mathbf{x}(n-P+1)] \\ \vdots & \ddots & \vdots \\ \mathbf{x}^T(n-P+1)[\mathbf{g}(n-1)\odot\mathbf{x}(n)] & \cdots & \delta + \mathbf{x}^T(n-P+1)[\mathbf{g}(n-1)\odot\mathbf{x}(n-P+1)] \end{bmatrix}.
\end{aligned}
\tag{9.27}
$$

It can be noticed that the matrix $\mathbf{M}(n)$ is symmetric but does not have a time-shift character. Consequently, we need to compute all its elements above (and including) the main diagonal, requiring $P^2L/2 + PL/2$ multiplications and $P^2(L-1)/2 + P(L+1)/2$ additions. On the other hand, according to (9.24) and (9.25), the matrix to be inverted in the MPAPA is

$$
\begin{aligned}
\mathbf{M}'(n) &= \delta \mathbf{I}_P + \mathbf{X}^T(n)\mathbf{P}'(n) \\
&= \delta \mathbf{I}_P + \begin{bmatrix} \mathbf{x}^T(n) \\ \vdots \\ \mathbf{x}^T(n-P+1) \end{bmatrix} \begin{bmatrix} \mathbf{g}(n-1)\odot\mathbf{x}(n) & \cdots & \mathbf{g}(n-P)\odot\mathbf{x}(n-P+1) \end{bmatrix} \\
&= \begin{bmatrix} \delta + \mathbf{x}^T(n)[\mathbf{g}(n-1)\odot\mathbf{x}(n)] & \cdots & \mathbf{x}^T(n)[\mathbf{g}(n-P)\odot\mathbf{x}(n-P+1)] \\ \vdots & \ddots & \vdots \\ \mathbf{x}^T(n-P+1)[\mathbf{g}(n-1)\odot\mathbf{x}(n)] & \cdots & \delta + \mathbf{x}^T(n-P+1)[\mathbf{g}(n-P)\odot\mathbf{x}(n-P+1)] \end{bmatrix}.
\end{aligned}
\tag{9.28}
$$

Clearly, this matrix is not symmetric, but has a time-shift property which allows us to rewrite (9.28) as

$$
\mathbf{M}'(n) = \begin{bmatrix} \delta + \mathbf{x}^T(n)[\mathbf{g}(n-1)\odot\mathbf{x}(n)] & \mathbf{x}^T(n)\mathbf{P}'_{-1}(n-1) \\ \mathbf{X}^T_{-1}(n-1)[\mathbf{g}(n-1)\odot\mathbf{x}(n)] & \mathbf{M}'_{P-1}(n-1) \end{bmatrix},
\tag{9.29}
$$

where the matrix $\mathbf{M}'_{P-1}(n-1)$ contains the first $P-1$ columns and $P-1$ rows of the matrix $\mathbf{M}'(n-1)$ [i.e., the top-left $(P-1)\times(P-1)$ submatrix of $\mathbf{M}'(n-1)$] and the matrix $\mathbf{X}_{-1}(n-1)$ contains the first $P-1$ columns of the matrix $\mathbf{X}(n-1)$. Consequently, only the first row and the first column of $\mathbf{M}'(n)$ need to be computed, requiring $2PL - L$ multiplications and $2P(L-1) - L + 2$ additions.

The classical PAPA and the MPAPA are summarized in Tables 9.1 and 9.2, respectively. The overall complexity of these algorithms depend on the method for obtaining the solution for $\mathbf{s}(n)$ (see Tables 9.1 and 9.2). However, computationally efficient methods for this task were recently proposed in (78). These methods are based on the dichotomous coordinate descent (DCD) iterations (77), i.e., a multiplication-free and division-free recursive technique that allows to efficiently solve a linear system of equations.

Table 9.1: The classical PAPA.	
Initialization:	$\hat{\mathbf{h}}(0) = \mathbf{0}$
Parameters:	$\delta = \text{cst} \cdot \sigma_x^2$
	$0 < \alpha < 2$
Error:	$\mathbf{e}(n) = \mathbf{d}(n) - \mathbf{X}^T(n)\hat{\mathbf{h}}(n-1)$
Compute:	$\mathbf{g}(n-1)$ (specific to the proportionate-type algorithm)
	$\mathbf{P}(n) = \begin{bmatrix} \mathbf{g}(n-1) \odot \mathbf{x}(n) & \cdots & \mathbf{g}(n-1) \odot \mathbf{x}(n-P+1) \end{bmatrix}$
	$\mathbf{M}(n) = \delta \mathbf{I}_P + \mathbf{X}^T(n)\mathbf{P}(n)$
Solve:	$\mathbf{M}(n)\mathbf{s}(n) = \mathbf{e}(n) \Longrightarrow \mathbf{s}(n)$
Update:	$\hat{\mathbf{h}}(n) = \hat{\mathbf{h}}(n-1) + \alpha\mathbf{P}(n)\mathbf{s}(n)$

Table 9.2: The "memory" PAPA (MPAPA).	
Initialization:	$\hat{\mathbf{h}}(0) = \mathbf{0}$
	$\mathbf{P}'_{-1}(0) = \mathbf{0}_{L \times (P-1)}$
	$\mathbf{M}'_{P-1}(0) = \delta \mathbf{I}_{P-1}$
Parameters:	$\delta = \text{cst} \cdot \sigma_x^2$
	$0 < \alpha < 2$
Error:	$\mathbf{e}(n) = \mathbf{d}(n) - \mathbf{X}^T(n)\hat{\mathbf{h}}(n-1)$
Compute:	$\mathbf{g}(n-1)$ (specific to the proportionate-type algorithm)
	$\mathbf{P}'(n) = \begin{bmatrix} \mathbf{g}(n-1) \odot \mathbf{x}(n) & \mathbf{P}'_{-1}(n-1) \end{bmatrix}$
	$\mathbf{P}'_{-1}(n) = $ the first $P-1$ columns of $\mathbf{P}'(n)$
	$\mathbf{M}'(n) = \begin{bmatrix} \delta + \mathbf{x}^T(n)\left[\mathbf{g}(n-1) \odot \mathbf{x}(n)\right] & \mathbf{x}^T(n)\mathbf{P}'_{-1}(n-1) \\ \mathbf{X}^T_{-1}(n-1)\left[\mathbf{g}(n-1) \odot \mathbf{x}(n)\right] & \mathbf{M}'_{P-1}(n-1) \end{bmatrix}$
	$\mathbf{M}'_{P-1}(n) = $ the top-left $(P-1) \times (P-1)$ submatrix of $\mathbf{M}'(n)$
	$\mathbf{X}_{-1}(n) = $ the first $P-1$ columns of $\mathbf{X}(n)$
Solve:	$\mathbf{M}'(n)\mathbf{s}(n) = \mathbf{e}(n) \Longrightarrow \mathbf{s}(n)$
Update:	$\hat{\mathbf{h}}(n) = \hat{\mathbf{h}}(n-1) + \alpha\mathbf{P}'(n)\mathbf{s}(n)$

A comparison between the computational complexities of the classical PAPA and MPAPA is given in Fig. 9.1, for a filter of length $L = 512$. We do not include here the complexities associated with the evaluation of $\mathbf{g}(n-1)$ (which depends on the proportionate-type algorithm) and the obtention of $\mathbf{s}(n)$. It is clear that the MPAPA is much more computationally efficient as compared to the classical version.

9.3   A VARIABLE STEP-SIZE VERSION

Similar to the case of PNLMS-type algorithms, the overall performance of PAPAs (in terms of convergence rate, tracking, and misadjustment) is governed by a step-size parameter. A constant

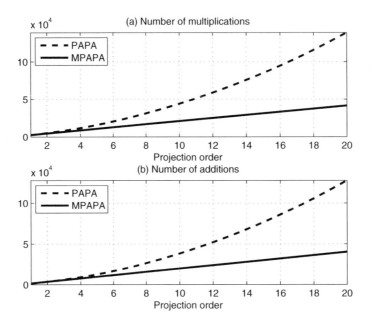

Figure 9.1: (a) Number of multiplications of the classical PAPA and MPAPA, as a function of the projection order. (b) Number of additions of the classical PAPA and MPAPA, as a function of the projection order. The length of the adaptive filter is $L = 512$.

value of this parameter leads to a compromise between the previous performances criteria. Thus, it is expected that a variable step-size (VSS) technique will further improve the behavior of PAPAs. In the following, we will derive a VSS-MPAPA by combining the MPAPA presented in the previous section with a recently developed VSS technique which was proven to be efficient for APAs (55).

For convenience of the derivation, we start by neglecting the regularization matrix $\delta \mathbf{I}_P$ in (9.25), so that the update of the MPAPA becomes

$$\hat{\mathbf{h}}(n) = \hat{\mathbf{h}}(n-1) + \alpha \mathbf{P}'(n) \left[\mathbf{X}^T(n) \mathbf{P}'(n) \right]^{-1} \mathbf{e}(n). \tag{9.30}$$

Next, let us rewrite the previous update as

$$\hat{\mathbf{h}}(n) = \hat{\mathbf{h}}(n-1) + \mathbf{P}'(n) \left[\mathbf{X}^T(n) \mathbf{P}'(n) \right]^{-1} \mathbf{D}_\alpha(n) \mathbf{e}(n), \tag{9.31}$$

where

$$\mathbf{D}_\alpha(n) = \mathrm{diag}\left[\begin{array}{cccc} \alpha_0(n) & \alpha_1(n) & \cdots & \alpha_{P-1}(n) \end{array} \right] \tag{9.32}$$

is a $P \times P$ diagonal matrix. It is clear that (9.30) is obtained when $\alpha_0(n) = \alpha_1(n) = \ldots = \alpha_{P-1}(n) = \alpha$. Defining the a posteriori error vector

$$\boldsymbol{\varepsilon}(n) = \mathbf{d}(n) - \mathbf{X}^T(n)\,\hat{\mathbf{h}}(n) \tag{9.33}$$

and using (9.31) and (9.1) in (9.33), we get

$$
\begin{aligned}
\boldsymbol{\varepsilon}(n) &= \mathbf{d}(n) - \mathbf{X}^T(n)\left\{\hat{\mathbf{h}}(n-1) + \mathbf{P}'(n)\left[\mathbf{X}^T(n)\,\mathbf{P}'(n)\right]^{-1}\mathbf{D}_\alpha(n)\,\mathbf{e}(n)\right\} \\
&= \mathbf{d}(n) - \mathbf{X}^T(n)\,\hat{\mathbf{h}}(n-1) - \mathbf{X}^T(n)\,\mathbf{P}'(n)\left[\mathbf{X}^T(n)\,\mathbf{P}'(n)\right]^{-1}\mathbf{D}_\alpha(n)\,\mathbf{e}(n) \\
&= \left[\mathbf{I}_P - \mathbf{D}_\alpha(n)\right]\mathbf{e}(n).
\end{aligned}
\tag{9.34}
$$

Let us assume that we deal with a system identification problem (33), so that the reference signal vector is obtained as

$$\mathbf{d}(n) = \mathbf{X}^T(n)\,\mathbf{h} + \mathbf{w}(n), \tag{9.35}$$

where $\mathbf{h}$ (of length L) denotes the impulse response vector of the unknown system and

$$\mathbf{w}(n) = \begin{bmatrix} w(n) & w(n-1) & \cdots & w(n-P+1) \end{bmatrix}^T$$

is a vector containing the last P samples of the system noise. In consistence with the basic idea of the APA, it can be imposed to cancel P a posteriori errors, i.e., $\boldsymbol{\varepsilon}(n) = \mathbf{0}_{P\times 1}$, where $\mathbf{0}_{P\times 1}$ denotes a column vector with all its P elements equal to zeros. Assuming that $\mathbf{e}(n) \neq \mathbf{0}_{P\times 1}$, it results from (9.34) that $\mathbf{D}_\alpha(n) = \mathbf{I}_P$. This corresponds to the update (9.30), with the step size $\alpha = 1$. In the absence of the system noise, i.e., $w(n) = 0$, we deal with an ideal "system identification" configuration. In this case, the value of the step size $\alpha = 1$ makes sense because it leads to the best performance (54). However, in most of real-world system identification scenarios (e.g., like echo cancellation) the existence of the system noise cannot be omitted, so that a more reasonable condition to impose is $\boldsymbol{\varepsilon}(n) = \mathbf{w}(n)$, i.e., to recover the system noise from the error of the adaptive filter. Thus, taking (9.34) into account, it results that

$$
\begin{aligned}
\varepsilon_p(n) &= \left[1 - \alpha_p(n)\right]e_p(n) \\
&= w(n-p),
\end{aligned}
\tag{9.36}
$$

where $\varepsilon_p(n)$ and $e_p(n)$ denote the $(p+1)$th elements of the vectors $\boldsymbol{\varepsilon}(n)$ and $\mathbf{e}(n)$, with $p = 0, 1, \ldots, P-1$. Our goal is to find an expression for the step-size parameter $\alpha_p(n)$ such that

$$E\left[\varepsilon_p^2(n)\right] = E\left[w^2(n-p)\right]. \tag{9.37}$$

Squaring (9.36) and taking the expectations it results:

$$\left[1 - \alpha_p(n)\right]^2 E\left[e_p^2(n)\right] = E\left[w^2(n-p)\right]. \tag{9.38}$$

By solving the quadratic equation (9.38), two solutions can be obtained. However, a value of the step size between 0 and 1 is preferable over the one between 1 and 2 [even if both solutions are stable but the former has less steady-state MSE with the same convergence speed (64)], so that it is reasonable to choose

$$\alpha_p(n) = 1 - \sqrt{\frac{E\left[w^2(n-p)\right]}{E\left[e_p^2(n)\right]}}. \tag{9.39}$$

From a practical point of view, (9.39) has to be evaluated in terms of power estimates as

$$\alpha_p(n) = 1 - \frac{\hat{\sigma}_w(n-p)}{\hat{\sigma}_{e_p}(n)}. \tag{9.40}$$

The variable in the denominator can be computed in a recursive manner, i.e.,

$$\hat{\sigma}_{e_p}^2(n) = \lambda\hat{\sigma}_{e_p}^2(n-1) + (1-\lambda)e_p^2(n), \tag{9.41}$$

where λ is an exponential window [its value is chosen as $\lambda = 1 - 1/(KL)$, with $K \geq 2$] and the initial value is $\hat{\sigma}_{e_p}^2(0) = 0$.

However, the main problem remains the estimation of the system noise power from the numerator of (9.40). In order to analyze this aspect, let us consider an echo cancellation scenario, which is the main application related to the proportionate-type algorithms. First, it is known that in the single-talk case, the near-end signal consists only of the background noise. Its power could be estimated during silences (and it can be assumed constant), so that (9.40) becomes

$$\alpha_p(n) = 1 - \frac{\hat{\sigma}_w}{\hat{\sigma}_{e_p}(n)}. \tag{9.42}$$

For a value of the projection order $P = 1$, the VSS-PNLMS algorithm from Section 8.2, Chapter 8 is obtained. For $P > 1$, a VSS-MPAPA can be derived, by computing (9.42) for $p = 0, 1, \ldots, P - 1$, then using a step-size matrix like in (9.32), and updating the filter coefficients according to (9.31). Nevertheless, the background noise can be time-variant, so that the power of the background noise should be periodically estimated. Moreover, when the background noise changes between two consecutive estimations or during the near-end speech, its new power estimate will not be available immediately; consequently, until the next estimation period of the background noise, the algorithm behavior will be disturbed. Second, in the double-talk case, the near-end signal consists of both the background noise and the near-end speech. It is very difficult to obtain an accurate estimate for the power of this combined signal, considering especially the non-stationary character of the speech signal.

In order to overcome these issues, let us consider the approach proposed in (55), which provides a simple but practical way to evaluate the numerator in (9.40). The reference signal can be expressed as

$$d(n) = y(n) + w(n), \tag{9.43}$$

where $y(n) = \mathbf{h}^T \mathbf{x}(n)$ is the output of the unknown system. Squaring (9.43) and taking the expectation of both sides [assuming that $y(n)$ and $w(n)$ are uncorrelated] it results that

$$E\left[d^2(n)\right] = E\left[y^2(n)\right] + E\left[w^2(n)\right], \tag{9.44}$$

so that,

$$E\left[w^2(n)\right] = E\left[d^2(n)\right] - E\left[y^2(n)\right]. \tag{9.45}$$

Supposing that the adaptive filter has converged to a certain degree, it can be considered that

$$E\left[y^2(n)\right] \approx E\left[\hat{y}^2(n)\right], \tag{9.46}$$

where $\hat{y}(n) = \hat{\mathbf{h}}^T(n-1)\mathbf{x}(n)$ is the output of the adaptive filter. Consequently,

$$E\left[w^2(n)\right] \approx E\left[d^2(n)\right] - E\left[\hat{y}^2(n)\right], \tag{9.47}$$

or in terms of power estimates

$$\hat{\sigma}_w^2(n) \approx \hat{\sigma}_d^2(n) - \hat{\sigma}_{\hat{y}}^2(n). \tag{9.48}$$

For the single-talk case, when only the background noise is present at the near-end, an estimate of its power is obtained using the right-hand term in (9.48). This expression holds even if the level of the background noise changes, so that there is no need for the estimation of this parameter during silences. For the double-talk case, when the near-end speech is present (assuming that it is uncorrelated with the background noise), the right-hand term in (9.48) also provides a power estimate of the near-end signal. More importantly, this term depends only on the signals that are available within the application, i.e., the reference signal, $d(n)$, and the output of the adaptive filter, $\hat{y}(n)$. Based on these findings, (9.40) can be rewritten as

$$\alpha_p(n) = 1 - \frac{\sqrt{\hat{\sigma}_d^2(n-p) - \hat{\sigma}_{\hat{y}}^2(n-p)}}{\hat{\sigma}_{e_p}(n)}, \quad p = 0, 1, \ldots, P-1. \tag{9.49}$$

As compared to (9.42), the previous relation is more suitable in practice. It should be noted that both terms from the numerator on the right-hand side of (9.49) can be evaluated using a recursive procedure similar to (9.41).

Finally, some practical issues need to be addressed. First, a very small positive number ϵ should be added to the denominator in (9.49) to avoid division by zero. Second, under our assumptions, we have $E\left[d^2(n-p)\right] \geq E\left[\hat{y}^2(n-p)\right]$ and $E\left[d^2(n-p)\right] - E\left[\hat{y}^2(n-p)\right] \approx E\left[e_p^2(n)\right]$. Nevertheless, the power estimates of these parameters could lead to some deviations from the previous

$$\text{Table 9.3: The variable step-size MPAPA (VSS-MPAPA).}$$

Initialization:	$\hat{\mathbf{h}}(0) = \mathbf{0}$
	$\mathbf{P}'_{-1}(0) = \mathbf{0}_{L \times (P-1)}$
	$\mathbf{M}'_{P-1}(0) = \delta \mathbf{I}_{P-1}$
	$\hat{\sigma}_d^2(n) = 0, \hat{\sigma}_{\hat{y}}^2(n) = 0, \text{ for } n \leq 0$
	$\hat{\sigma}_{e_p}(0) = 0, \text{ for } p = 0, 1, \ldots, P-1$
Parameters:	$\delta = \text{cst} \cdot \sigma_x^2$
	$\lambda = 1 - 1/(KL), \text{ exponential window with } K \geq 2$
	$\epsilon > 0, \text{ very small number to avoid division by zero}$
Error:	$\mathbf{e}(n) = \mathbf{d}(n) - \mathbf{X}^T(n)\hat{\mathbf{h}}(n-1)$
Compute:	$\hat{y}(n) = \hat{\mathbf{h}}^T(n-1)\mathbf{x}(n)$
	$\mathbf{g}(n-1) \text{ (specific to the proportionate-type algorithm)}$
	$\mathbf{P}'(n) = \begin{bmatrix} \mathbf{g}(n-1) \odot \mathbf{x}(n) & \mathbf{P}'_{-1}(n-1) \end{bmatrix}$
	$\mathbf{P}'_{-1}(n) = \text{the first } P-1 \text{ columns of } \mathbf{P}'(n)$
	$\mathbf{M}'(n) = \begin{bmatrix} \delta + \mathbf{x}^T(n)\left[\mathbf{g}(n-1) \odot \mathbf{x}(n)\right] & \mathbf{x}^T(n)\mathbf{P}'_{-1}(n-1) \\ \mathbf{X}^T_{-1}(n-1)\left[\mathbf{g}(n-1) \odot \mathbf{x}(n)\right] & \mathbf{M}'_{P-1}(n-1) \end{bmatrix}$
	$\mathbf{M}'_{P-1}(n) = \text{the top-left } (P-1) \times (P-1) \text{ submatrix of } \mathbf{M}'(n)$
	$\mathbf{X}_{-1}(n) = \text{the first } P-1 \text{ columns of } \mathbf{X}(n)$
Updates:	$\hat{\sigma}_{e_p}^2(n) = \lambda\hat{\sigma}_{e_p}^2(n-1) + (1-\lambda)e_p^2(n), \text{ for } p = 0, 1, \ldots, P-1$
	$\hat{\sigma}_d^2(n) = \lambda\hat{\sigma}_d^2(n-1) + (1-\lambda)d^2(n)$
	$\hat{\sigma}_{\hat{y}}^2(n) = \lambda\hat{\sigma}_{\hat{y}}^2(n-1) + (1-\lambda)\hat{y}^2(n)$
	$\alpha_p(n) = \left| 1 - \dfrac{\sqrt{|\hat{\sigma}_d^2(n-p) - \hat{\sigma}_{\hat{y}}^2(n-p)|}}{\epsilon + \hat{\sigma}_{e_p}(n)} \right|, \text{ for } p = 0, 1, \ldots, P-1$
	$\mathbf{D}_\alpha(n) = \text{diag}\begin{bmatrix} \alpha_0(n) & \alpha_1(n) & \cdots & \alpha_{P-1}(n) \end{bmatrix}$
Solve:	$\mathbf{q}(n) = \mathbf{D}_\alpha(n)\mathbf{e}(n)$
	$\mathbf{M}'(n)\mathbf{s}(n) = \mathbf{q}(n) \Longrightarrow \mathbf{s}(n)$
Filter update:	$\hat{\mathbf{h}}(n) = \hat{\mathbf{h}}(n-1) + \mathbf{P}'(n)\mathbf{s}(n)$

theoretical conditions, so that we will take the absolute values in (9.49). Hence, the final step-size formula is rewritten as

$$\alpha_p(n) = \left| 1 - \frac{\sqrt{|\hat{\sigma}_d^2(n-p) - \hat{\sigma}_{\hat{y}}^2(n-p)|}}{\epsilon + \hat{\sigma}_{e_p}(n)} \right|, \quad p = 0, 1, \ldots, P-1. \tag{9.50}$$

Summarizing, the resulted VSS-MPAPA is listed in Table 9.3.

Since it is based on the assumption that the adaptive filter coefficients have converged to a certain degree, this VSS-MPAPA could experience a slower initial convergence rate and a slower tracking capability as compared to the MPAPA, because (9.46) is biased in these situations. Concerning the initial convergence rate, we could start the proposed algorithm using a regular MPAPA [i.e., using a (high) fixed step size] in the first N iterations, with $N \geq L$. Nevertheless, even if we do not use this "trick," the experimental results will prove that the performance degradation is not very significant (especially when the value of the projection order is increased).

CHAPTER 10

Experimental Study

Throughout the chapters of this book, we have reviewed some of the most important sparse adaptive filters. However, many other versions and combinations of these algorithms can be found in the literature. For this reason, it would be extremely difficult (and sometimes even confusing) to compare and analyze all these possible algorithms. Consequently, the goal of this chapter is to outline by means of simulations the most important features of the main proportionate-type algorithms described in this book, in order to emphasize their capabilities in different practical echo cancellation scenarios. To facilitate the flow of the experiments reported in this chapter, we will follow the structure of the book, by analyzing first the basic proportionate-type NLMS adaptive filters (reviewed in Chapter 5 and 6), then the mu-law and other recently developed PNLMS-type algorithms (described in Chapter 7), followed by the variable step-size versions (developed in Chapter 8), and finally the proportionate-type APAs (presented in Chapter 9).

10.1 EXPERIMENTAL CONDITIONS

Experiments were performed in the context of echo cancellation since this is the main application of sparse adaptive filters. Two echo paths were used (see Fig. 10.1), having different sparseness degree. The first one [Fig. 10.1(a)] is a network echo path from G168 Recommendation (79); its impulse response can be considered to be very sparse since the associated sparseness measure is $\xi_{12}(\mathbf{h}) = 0.8970$ [see (2.26) in Section 2.4, Chapter 2]. The second one [Fig. 10.1(b)] is a measured acoustic echo path, which seems less sparse from the figure and is confirmed by the sparseness measure, i.e., $\xi_{12}(\mathbf{h}) = 0.6131$. Both impulse responses have 512 coefficients, using a sampling rate of 8 kHz. All adaptive filters used in the experiments have the same length, i.e., $L = 512$.

The far-end signal (i.e., the input signal) is either a white Gaussian signal or a speech sequence. The output of the echo path is corrupted by an independent white Gaussian noise (i.e., the background noise at the near-end) with 30 dB echo-to-noise ratio (ENR), see Chapter 3. Only in one experiment (reported in Fig. 10.17), a case with ENR = 5 dB is also evaluated. Most of the simulations, except for the last one reported in Fig. 10.20, are performed in the single-talk case. In order to evaluate the tracking capabilities of the algorithms, an echo path change scenario is simulated in some experiments, by shifting the impulse response to the right by 12 samples.

For a fair comparison, the normalized step-size parameter was set to $\alpha = 0.2$ for all the algorithms, except in the experiments with a variable step-size (VSS) parameter. Also, the regularization parameters of the classical NLMS algorithm and APA were fixed to $\delta_{\text{NLMS}} = \delta_{\text{APA}} = 20\sigma_x^2$, where σ_x^2 is the input signal variance; the regularization parameters of the other algorithms are chosen as

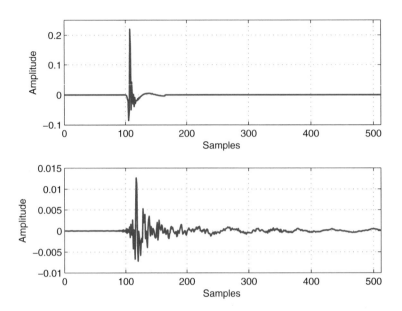

Figure 10.1: Impulse responses used in simulations. (a) Network echo path, $\xi_{12}(\mathbf{h}) = 0.8970$. (b) Acoustic echo path, $\xi_{12}(\mathbf{h}) = 0.6131$.

a function of this value. For the VSS algorithms, the forgetting factor λ uses $K = 2$ for the white Gaussian input signal and $K = 6$ for the speech input signal.

The performance measure used in most of the experiments (except for the last one) is the normalized misalignment (in dB) evaluated according to (3.14) (see Chapter 3). The last simulation performed in a double-talk scenario (Fig. 10.20) is evaluated in terms of the ERLE (see Chapter 3). In all the experiments, the results are averaged over 20 independent trials.

10.2   IPNLMS VERSUS PNLMS

The algorithms evaluated in this section could be considered as the classical ones in the field because they are the benchmarks to which most of the new developed proportionate-type algorithms are compared. We mainly refer here to the PNLMS (18) and improved PNLMS (IPNLMS) algorithms (6). These adaptive filters use some initialization parameters controlling their overall performance. There are many works that analyze the influence of these parameters over the performance of the algorithms. However, typical values of these parameters (recommended in most of the practical applications, when no a priori information is available) are

- for the PNLMS algorithm (see Table 5.1 in Chapter 5): $\delta_{\mathrm{p}} = 0.01$ and $\varrho = 5/L$;

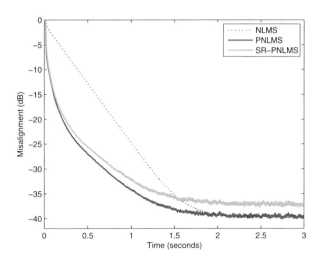

Figure 10.2: Misalignment of the NLMS, PNLMS, and SR-PNLMS algorithms with a white Gaussian input signal and impulse response of Fig. 10.1(a).

- for the IPNLMS algorithm (see Table 5.2 in Chapter 5): $\kappa = 0$ or -0.5.

In our experiments, we use $\kappa = 0$ because it seems to be a "fair" choice (see the discussion in Section 5.4, Chapter 5). Also, the choice of the regularization parameter is important for these algorithms. Accordingly, we set $\delta_{\text{PNLMS}} = \delta_{\text{NLMS}}/L$ and $\delta_{\text{IPNLMS}} = \delta_{\text{NLMS}}(1 - \kappa)/(2L)$.

First, we compare the PNLMS (18) and NLMS algorithms in order to outline the capabilities of the proportionate concept. Figure 10.2 presents the misalignment of these algorithms when identifying the very sparse impulse response from Fig. 10.1(a) and using a white Gaussian input signal. Also, the signed-regressor PNLMS (SR-PNLMS) algorithm was included for comparison (see Section 5.3 in Chapter 5); this algorithm can be viewed as a "cheaper" version of the PNLMS algorithm (in terms of computational complexity). It is clear from Fig. 10.2 that the PNLMS algorithm outperforms the NLMS in terms of convergence rate; the SR-PNLMS algorithm performs similarly to the PNLMS, but achieves a higher level of misalignment.

However, the performance of the PNLMS algorithm degrades when the impulse response is not so sparse. Figure 10.3 supports this aspect by comparing the algorithms with the acoustic echo path from Fig. 10.1(b). The PNLMS++ algorithm (25) was also included for comparison; let us remember that this algorithm alternates between NLMS and PNLMS iterations, thus becoming less sensitive to the sparseness degree of the impulse response (see Section 5.2 in Chapter 5). It can be noticed from Fig. 10.3 that the PNLMS algorithm becomes even slower than the NLMS (after the initial convergence phase), while the PNLMS++ algorithm performs well in this situation, achieving the fastest convergence rate with a slightly higher misalignment.

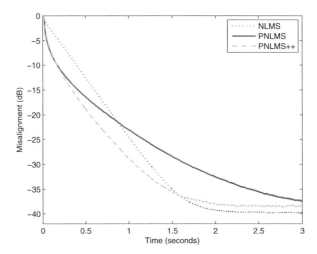

Figure 10.3: Misalignment of the NLMS, PNLMS, and PNLMS++ algorithms with a white Gaussian input signal and impulse response of Fig. 10.1(b).

In the following simulations, we include for comparison the IPNLMS algorithm (6). The main feature of this algorithm is its robustness to the sparseness degree of the echo path. As a reference, the exponentiated gradient algorithm with positive and negative weights (EG± algorithm) (41) was included for comparison. It was shown in Chapter 6 that this algorithm is connected with the IPNLMS. However, the performance of this algorithm depends on the parameter $\mathcal{H}$ (see Table 6.1 in Chapter 6) related to the ℓ_1 norm of the impulse response to be identified. Consequently, the choice of this parameter is critical when no a priori information is available, making the EG± algorithm difficult to use in practice. In our simulations, we choose $\mathcal{H} = 2 \|\mathbf{h}\|_1$, where $\mathbf{h}$ denotes the true impulse response of the echo path. The other initialization parameters of the EG± algorithm were set to $c = 1$ and $\delta_{\mathrm{EG\pm}} = \delta_{\mathrm{NLMS}}$.

Figure 10.4 compares the misalignment of the PNLMS, EG±, and IPNLMS algorithms with the network echo path from Fig. 10.1(a); the input signal is white and Gaussian. It can be noticed that the PNLMS algorithm performs similarly to the EG± algorithm, while the IPNLMS algorithm is slightly better in terms of convergence rate. The advantage of the IPNLMS algorithm becomes more apparent when identifying less sparse impulse responses. The previous simulation is repeated in Fig. 10.5 but using the acoustic echo path from Fig. 10.1(b). In this situation, it can be noticed that the IPNLMS outperforms, by far, the PNLMS algorithm in terms of convergence rate. The EG± algorithm performs similarly to the IPNLMS, confirming that these algorithms are related.

The regular IPNLMS algorithm (6) uses the ℓ_1 norm to exploit the sparsity of the impulse response that needs to identified; however, a better measure of sparseness is the ℓ_0 norm. The IPNLMS

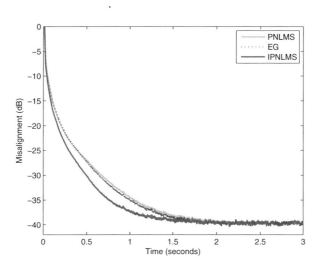

Figure 10.4: Misalignment of the PNLMS, EG±, and IPNLMS algorithms with a white Gaussian input signal and impulse response of Fig. 10.1(a).

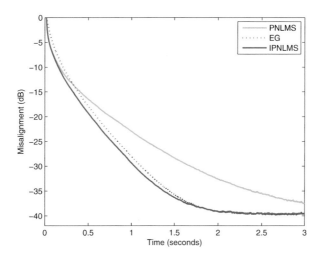

Figure 10.5: Misalignment of the PNLMS, EG±, and IPNLMS algorithms with a white Gaussian input signal and impulse response of Fig. 10.1(b).

with the ℓ_0 norm algorithm proposed in (57) exploits this idea (see Section 5.4 in Chapter 5). As compared to the regular IPNLMS, this algorithm introduces a new parameter, β_0, which is used in the practical evaluation of the ℓ_0 norm (as a continuous exponential function). The value of this

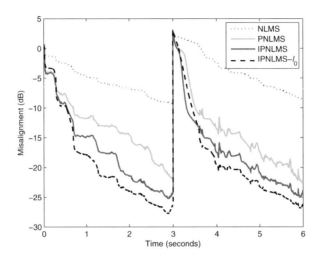

Figure 10.6: Misalignment of the NLMS, PNLMS, regular IPNLMS, and ℓ_0-norm IPNLMS algorithms with a speech input signal and impulse response of Fig. 10.1(a). The impulse response changes at time 3 seconds.

parameter depends on the sparseness of the impulse response; in our simulation, we set $\beta_0 = 50$. Figure 10.6 compares the misalignment of the NLMS, PNLMS, regular IPNLMS, and ℓ_0-norm IPNLMS algorithms in a more realistic scenario, i.e., using a speech signal as input and introducing an echo path change at time 3 seconds; the impulse response from Fig. 10.1(a) is used. It can be noticed that the IPNLMS with the ℓ_0 norm outperforms the other algorithms. In this case, it is interesting to notice that the improvement of the IPNLMS with the ℓ_0 norm over the IPNLMS algorithm is similar with the improvement of the IPNLMS over the PNLMS algorithm.

10.3   MPNLMS, SC-PNLMS, AND IAF-PNLMS

The mu-law PNLMS (MPNLMS) algorithm (14), (15) was derived following an "optimal" approach (based on the steepest-descent method), such that all the coefficients of the adaptive filter reach the ζ-vicinity of their optimal values after the same number of iterations, typically, $\zeta = 0.001$ (see Table 7.1 in Chapter 7). Different versions of the MPNLMS algorithm were proposed. One of the most recent ones is the adaptive MPNLMS (AMPNLMS) algorithm (71), which uses a variable parameter $\zeta(n)$ instead of a constant one. Figure 10.7 compares the misalignment of the PNLMS, IPNLMS, MPNLMS, and AMPNLMS algorithms when identifying the network echo path from Fig. 10.1(a). The input signal is white, Gaussian, and the echo path changes at time 1.5 seconds. The regularization parameters of the MPNLMS algorithms are $\delta_{\text{MPNLMS}} = \delta_{\text{AMPNLMS}} = \delta_{\text{NLMS}}$. The other parameters specific to the AMPNLMS algorithm were set to $\lambda = 0.99$ and $\varphi = 0.001$

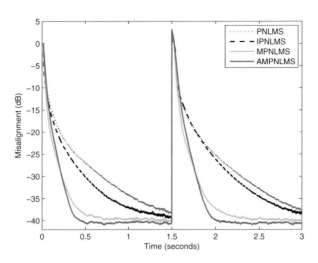

Figure 10.7: Misalignment of the PNLMS, IPNLMS, MPNLMS, and AMPNLMS algorithms with a white Gaussian input signal and impulse response of Fig. 10.1(a). The impulse response changes at time 1.5 seconds.

(see Chapter 7). It can be noticed from Fig. 10.7 that the MPNLMS algorithms outperform both PNLMS and IPNLMS in this case. This is due to the "optimal" approach used in the derivation of the MPNLMS algorithm.

In Fig. 10.8, the algorithms are compared using a speech sequence as input. The individual activation factor PNLMS (IAF-PNLMS) algorithm proposed in (68) was also included for comparison (see Chapter 7); the regularization parameter of this algorithm is set to $\delta_{\text{IAF-PNLMS}} = \delta_{\text{PNLMS}}$. According to this simulation, the IAF-PNLMS algorithm performs similarly to the IPNLMS, while the MPNLMS outperforms all the other algorithms. However, both MPNLMS and IAF-PNLMS algorithms were developed following the equations of the PNLMS algorithm, thus being suitable when identifying very sparse impulse responses. In Fig. 10.9, the algorithms are compared when identifying the impulse response from Fig. 10.1(b); the input signal is white and Gaussian. It can be noticed that the performance of the MPNLMS and IAF-PNLMS algorithms significantly degrades. In this case, the advantage of the AMPNLMS algorithm becomes more apparent; it performs similarly to the IPNLMS, outperforming the other algorithms.

Another recent solution to overcome the sensitivity of the PNLMS algorithm to the sparseness degree of the echo path was proposed in (45). The sparseness-controlled PNLMS (SC-PNLMS) algorithm incorporates an estimated sparseness measure, in order to improve the robustness to varying levels of sparseness of impulse responses (see Chapter 7). In the next simulation, the parameters of this algorithm were set to $\delta_{\text{SC-PNLMS}} = \delta_{\text{PNLMS}}$ and $\rho = 6$ (see Table 7.2 in Chapter 7). Figure 10.10 compares the performance of the PNLMS, IPNLMS, and SC-PNLMS algorithms when

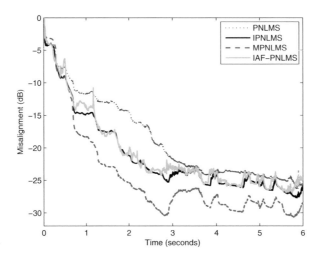

Figure 10.8: Misalignment of the PNLMS, IPNLMS, MPNLMS, and IAF-PNLMS algorithms with a speech input signal and impulse response of Fig. 10.1(a).

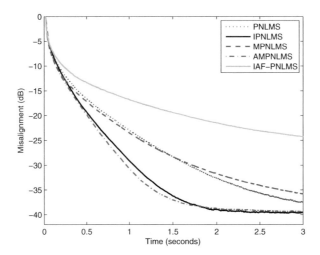

Figure 10.9: Misalignment of the PNLMS, IPNLMS, MPNLMS, AMPNLMS, and IAF-PNLMS algorithms with a white Gaussian input signal and impulse response of Fig. 10.1(b).

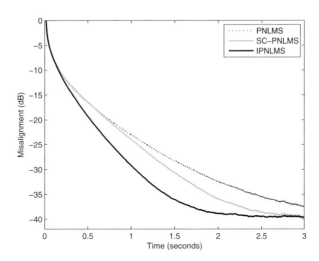

Figure 10.10: Misalignment of the PNLMS, SC-PNLMS, and IPNLMS algorithms with a white Gaussian input signal and impulse response of Fig. 10.1(b).

identifying the acoustic impulse response from Fig. 10.1(b); the input signal is white and Gaussian. It can be noticed that the SC-PNLMS algorithm outperforms the PNLMS, achieving a faster convergence. Nevertheless, the IPNLMS algorithm still performs better in terms of convergence rate.

10.4   VSS-IPNLMS

The VSS approach presented in Chapter 8 can be combined with any PNLMS-type algorithm. However, among the previous discussed algorithms, we choose to combine it with the IPNLMS. There are three main reasons behind this choice: 1) the IPNLMS algorithm is easy to implement and to control in practice, 2) it is robust to the sparseness degree of the echo path, and 3) it outperforms most of the PNLMS-type algorithms. As compared to the IPNLMS algorithm, the resulted VSS-IPNLMS needs an estimate of the background noise power, σ_w^2. In echo cancellation, it can be easily estimated during silences. Also, there are other practical methods to evaluate this parameter (56). In our simulations, we assume that the value of σ_w^2 is known.

Figure 10.11 compares the performance of the IPNLMS using two different normalized step-size parameters (i.e., $\alpha = 0.1$ and $\alpha = 1$) with the VSS-IPNLMS algorithm, with the impulse response from Fig. 10.1(a), and using a white Gaussian input signal. Clearly, the VSS-IPNLMS outperforms the IPNLMS algorithm, achieving similar convergence rate with the fastest IPNLMS (i.e., with the largest normalized step-size parameter) and the final misalignment of the IPNLMS with a much smaller normalized step-size parameter.

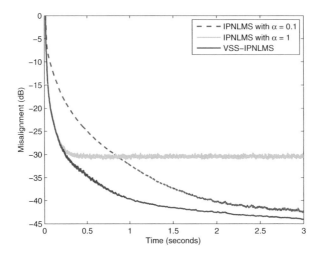

Figure 10.11: Misalignment of the IPNLMS algorithm with two different values of the normalized step-size parameter ($\alpha = 0.1$ and $\alpha = 1$) and misalignment of the VSS-IPNLMS algorithm, with a white Gaussian input signal and impulse response of Fig. 10.1(a).

It is also interesting to compare the performance of the VSS-IPNLMS algorithm with its non-proportionate counterpart, i.e., the non-parametric VSS-NLMS (NPVSS-NLMS) algorithm (9) presented in Section 4.4, Chapter 4. In Fig. 10.12, these two algorithms are evaluated when identifying the impulse response from Fig. 10.1(a), using speech as input, and introducing an echo path change at time 3 seconds. According to this simulation, it is obvious that the VSS-IPNLMS algorithm performs much better than the NPVSS-NLMS. However, this difference becomes less apparent when identifying a less sparse impulse response. The previous experiment is repeated in Fig. 10.13, but using the acoustic echo path from Fig. 10.1(b). In this situation, it can be noticed that the NPVSS-NLMS performs closely to the IPNLMS algorithm with the normalized step-size parameter $\alpha = 0.2$.

10.5   PAPAS

The idea of the IPNLMS algorithm was straightforwardly extended to the APA, thus resulting the improved proportionate APA (IPAPA) (36). It is known that the IPNLMS algorithm always outperforms the NLMS. Consequently, a similar advantage of the IPAPA over the APA is expected. This is certified by the experiments reported in Fig. 10.14 [using the impulse response from Fig. 10.1(a)] and in Fig. 10.15 [with the impulse response from Fig. 10.1(b)]; the input signal is white and Gaussian, and the echo path changes at time 1.5 seconds. The regularization parameter of the IPAPA is $\delta_{\text{IPAPA}} = \delta_{\text{IPNLMS}}$ and the projection order is $P = 2$. It can be noticed from Fig. 10.14 that the

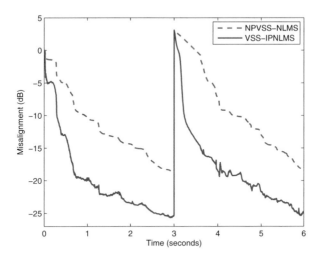

Figure 10.12: Misalignment of the NPVSS-NLMS and VSS-IPNLMS algorithms with a speech input signal and impulse response of Fig. 10.1(a). The impulse response changes at time 3 seconds.

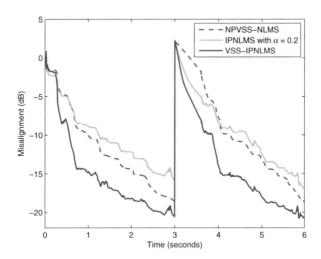

Figure 10.13: Misalignment of the NPVSS-NLMS, IPNLMS (with the normalized step-size parameter $\alpha = 0.2$), and VSS-IPNLMS algorithms with a speech input signal and impulse response of Fig. 10.1(b). The impulse response changes at time 3 seconds.

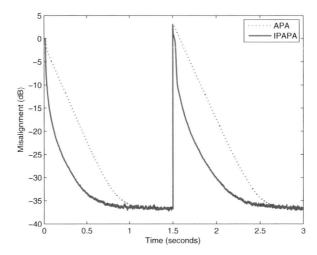

Figure 10.14: Misalignment of the APA and IPAPA for $P = 2$ with a white Gaussian input signal and impulse response of Fig. 10.1(a). The impulse response changes at time 1.5 seconds.

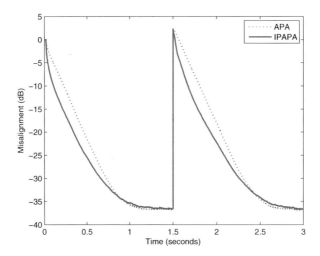

Figure 10.15: Misalignment of the APA and IPAPA for $P = 2$ with a white Gaussian input signal and impulse response of Fig. 10.1(b). The impulse response changes at time 1.5 seconds.

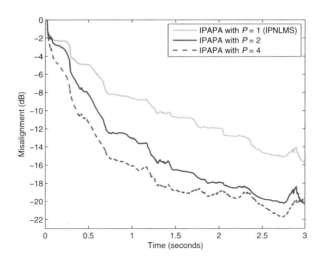

Figure 10.16: Misalignment of the IPAPA with different projection orders [$P = 1$ (IPNLMS), $P = 2$, and $P = 4$] with a speech input signal and impulse response of Fig. 10.1(b).

IPAPA outperforms, by far, the APA when identifying very sparse impulse responses. According to Fig. 10.15, the differences between the algorithms become less apparent in the case of less sparse echo paths, which is also expected.

The influence of the projection order over the performance of the IPAPA is evaluated in Fig. 10.16. The input signal is speech and the impulse response from Fig. 10.1(b) is used. The projection orders used in this experiment are $P = 1$ (equivalent to the IPNLMS algorithm), $P = 2$, and $P = 4$. It can be noticed that there is a significant performance improvement of the IPAPA with $P = 2$ over the case $P = 1$. However, the improvement is less significant when the projection order P increases, as compared to the case with $P = 2$. These results are in consistence with the well-known behavior of the APA.

Similar to the VSS technique discussed in the previous section, the recently proposed proportionate-type APA developed in Chapter 9 can be combined with any proportionate factors. Motivated by the same reasons as before, we chose to combine it with the proportionate parameters of the IPNLMS algorithm, thus resulting a "memory" IPAPA (MIPAPA) (58). In Fig. 10.17, the performance of the MIPAPA is compared to IPAPA for a projection order $P = 8$; the regularization parameter is $\delta_{\text{MIPAPA}} = \delta_{\text{IPAPA}}$. The impulse response from Fig. 10.1(a) is used, the input signal is white, Gaussian, and the echo path changes at time 0.5 seconds. Also, different values of the ENR are used, i.e., 30 dB and 5 dB. It can be noticed that the MIPAPA outperforms its counterpart providing both faster tracking and lower misadjustment, even for low ENR values. Also, we should remember that the MIPAPA is more efficient to implement than the IPAPA (see Chapter 9).

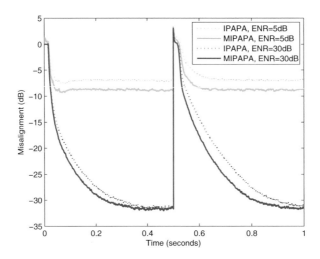

Figure 10.17: Misalignment of the IPAPA and MIPAPA for $P = 8$ and different ENRs (30 dB and 5 dB), with a white Gaussian input signal and impulse response of Fig. 10.1(a). The impulse response changes at time 0.5 seconds.

Finally, the combination between MIPAPA and the VSS technique developed in Chapter 9 is presented. In Fig. 10.18, the resulted VSS-MIPAPA is compared with the MIPAPA using two different step-size parameters, i.e., $\alpha = 1$ and $\alpha = 0.02$; the projection order is $P = 4$, the input signal is white and Gaussian, and the impulse response from Fig. 10.1(a) is used. It can be noticed that the VSS-MIPAPA has a convergence rate slightly slower that the MIPAPA with the largest step size, but it achieves a significant lower misalignment, which is close to the one obtained by the MIPAPA with a much smaller step size.

Next, the VSS-MIPAPA is compared with its non-proportional counterpart, i.e., the VSS-APA proposed in (55). The projection order is $P = 4$ and the input signal is speech; the impulse response from Fig. 10.1(a) is used and the echo path changes at time 2.5 seconds. The results are given in Fig. 10.19. It can be noticed that the VSS-MIPAPA outperforms the VSS-APA in terms of both convergence rate and misalignment.

The main feature of VSS algorithms is their robustness to double talk (55). The last experiment evaluates the performance of the MIPAPA (with $\alpha = 0.2$), VSS-APA, and VSS-MIPAPA in a double-talk scenario. The input signal is speech, the echo path from Fig. 10.1(a) is used and the near-end speech appears between times 1.25 and 2.5 seconds. A simple Geigel DTD is involved in this simulation; the threshold is equal to 0.5 and the hangover time is set to 240 samples (23). The results are given in Fig. 10.20 in terms of the ERLE. It can be noticed that the VSS-MIPAPA outperforms the other algorithms.

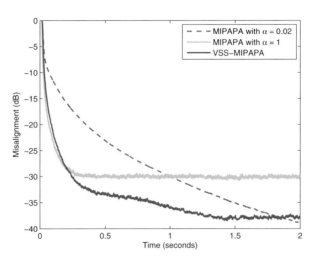

Figure 10.18: Misalignment of the MIPAPA with two different values of the step-size parameter ($\alpha = 1$ and $\alpha = 0.02$) and misalignment of the VSS-MIPAPA for $P = 4$, with a white Gaussian input signal and impulse response of Fig. 10.1(a).

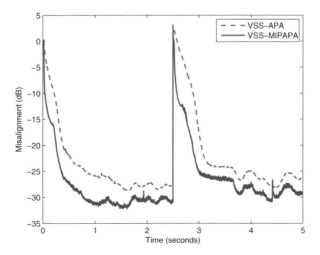

Figure 10.19: Misalignment of the VSS-APA and VSS-MIPAPA for $P = 4$, with a speech input signal and impulse response of Fig. 10.1(a). The impulse response changes at time 2.5 seconds.

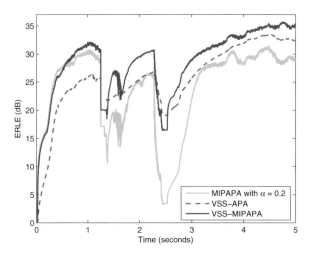

Figure 10.20: ERLE of the MIPAPA (with a step-size parameter $\alpha = 0.2$), VSS-APA, and VSS-MIPAPA for $P = 4$, with a speech input signal and impulse response of Fig. 10.1(a). Double talk appears between times 1.5 and 2.5 seconds; a Geigel DTD is used.

Bibliography

[1] T. Aboulnasr and K. Mayyas, "A robust variable step-size LMS-type algorithm: analysis and simulations," *IEEE Trans. Signal Processing*, vol. 45, pp. 631–639, Mar. 1997. DOI: 10.1109/78.558478 31

[2] M. G. Bellanger, *Adaptive Digital Filters and Signal Analysis*. New York: Marcel Dekker, 1987. 35

[3] J. Benesty, D. R. Morgan, and M. M. Sondhi, "A better understanding and an improved solution to the specific problems of stereophonic acoustic echo cancellation," *IEEE Trans. Speech, Audio Processing*, vol. 6, pp. 156–165, Mar. 1998. DOI: 10.1109/89.661474 16

[4] J. Benesty, D. R. Morgan, and J. H. Cho, "A new class of doubletalk detectors based on cross-correlation," *IEEE Trans. Speech, Audio Processing*, vol. 8, pp. 168–172, Mar. 2000. 3

[5] J. Benesty, T. Gänsler, D. R. Morgan, M. M. Sondhi, and S. L. Gay, *Advances in Network and Acoustic Echo Cancellation*. Berlin, Germany: Springer-Verlag, 2001. 1, 3, 16, 17, 40, 41, 47, 53

[6] J. Benesty and S. L. Gay, "An improved PNLMS algorithm," in *Proc. IEEE ICASSP*, 2002, pp. 1881–1884. 4, 42, 88, 90

[7] J. Benesty, Y. Huang, and D. R. Morgan, "On a class of exponentiated adaptive algorithms for the identification of sparse impulse responses," in *Adaptive Signal Processing–Applications to Real-World Problems*, J. Benesty and Y. Huang, Eds., Berlin: Springer, Chapter 1, pp. 1–22, 2003. 49, 52

[8] J. Benesty and Y. Huang, "The LMS, PNLMS, and exponentiated gradient algorithms," in *Proc. EUSIPCO*, 2004, pp. 721–724. 49, 52

[9] J. Benesty, H. Rey, L. Rey Vega, and S. Tressens, "A non-parametric VSS-NLMS algorithm," *IEEE Signal Processing Lett.*, vol. 13, pp. 581–584, Oct. 2006. DOI: 10.1109/LSP.2006.876323 3, 32, 33, 96

[10] N. J. Bershad, "On the optimum data non-linearity in LMS adaptation," *IEEE Trans. Acoust., Speech, Signal Processing*, vol. ASSP-34, pp. 69–76, Feb. 1986. DOI: 10.1109/TASSP.1986.1164798 35

[11] T. Claasen and W. Mecklenbrauker, "Comparison of the convergence of two algorithms for adaptive FIR digital filters," *IEEE Trans. Acoust., Speech, Signal Processing*, vol. ASSP-29, pp. 670–678, June 1981. DOI: 10.1109/TASSP.1981.1163596 35

[12] P. M. Clarkson, *Optimal and Adaptive Signal Processing*. London, UK: CRC, 1993. 35

[13] L. Couch, *Digital and Analog Communication Systems*. Fifth Edition, Englewood Cliffs, NJ: Prentice-Hall, 1997. 58

[14] H. Deng and M. Doroslovački, "Improving convergence of the PNLMS algorithm for sparse impulse response identification," *IEEE Signal Processing Lett.*, vol. 12, pp. 181–184, Mar. 2005. 5, 55, 58, 92

[15] H. Deng and M. Doroslovački, "Proportionate adaptive algorithms for network echo cancellation," *IEEE Trans. Signal Processing*, vol. 54, pp. 1794–1803, May 2006. DOI: 10.1109/TSP.2006.872533 5, 55, 56, 57, 58, 92

[16] D. Donoho, "Compressed sensing," *IEEE Trans. Information Theory*, vol. 52, pp. 1289–1306, Apr. 2006. DOI: 10.1109/TIT.2006.871582 7, 44

[17] D. L. Duttweiler, "A twelve-channel digital echo canceler," *IEEE Trans. Communications*, vol. 26, pp. 647–653, May 1978. DOI: 10.1109/TCOM.1978.1094133 3

[18] D. L. Duttweiler, "Proportionate normalized least-mean-squares adaptation in echo cancelers," *IEEE Trans. Speech, Audio Processing*, vol. 8, pp. 508–518, Sept. 2000. DOI: 10.1109/89.861368 4, 37, 39, 55, 60, 88, 89

[19] G. Enzer and P. Vary, "Robust and elegant, purely statistical adaptation of acoustic echo canceler and postfilter," in *Proc. IWAENC*, 2003, pp. 43–46. 3

[20] J. B. Evans, P. Xue, and B. Liu, "Analysis and implementation of variable step size adaptive algorithms," *IEEE Trans. Signal Processing*, vol. 41, pp. 2517–2535, Aug. 1993. DOI: 10.1109/78.229885 31

[21] A. Feuer and E. Weinstein, "Convergence analysis of LMS filters with uncorrelated Gaussian data," *IEEE Trans. Acoust., Speech, Signal Processing*, vol. ASSP-33, pp. 222–230, Feb. 1985. DOI: 10.1109/TASSP.1985.1164493 28

[22] T. Gaensler, M. Hansson, C.-J. Ivanson, and G. Salomonsson, "A double-talk detector based on coherence," *IEEE Trans. Communications*, vol. 44, pp. 1421–1427, Nov. 1996. DOI: 10.1109/26.544458 3

[23] T. Gaensler, S. L. Gay, M. M. Sondhi, and J. Benesty, "Double-talk robust fast converging algorithms for network echo cancellation," *IEEE Trans. Speech, Audio Processing*, vol. 8, pp. 656–663, Nov. 2000. DOI: 10.1109/89.876299 3, 74, 75, 100

[24] S. L. Gay and S. Tavathia, "The fast affine projection algorithm," in *Proc. IEEE ICASSP*, 1995, vol. 5, pp. 3023–3026. 2, 78

[25] S. L. Gay, "An efficient, fast converging adaptive filter for network echo cancellation," in *Proc. of Asilomar*, 1998, pp. 394–398. 40, 89

[26] S. L. Gay and S. C. Douglas, "Normalized natural gradient adaptive filtering for sparse and nonsparse systems," in *Proc. IEEE ICASSP*, 2002, pp. 1405–1408. 54

[27] A. Gersho, "Adaptive filtering with binary reinforcement," *IEEE Trans. Information Theory*, vol. IT-30, pp. 191–199, Mar. 1984. DOI: 10.1109/TIT.1984.1056890 34

[28] S. Gollamudi, S. Nagaraj, S. Kapoor, and Y.-F. Huang, "Set-membership filtering and a set-membership normalized LMS algorithm with an adaptive step size," *IEEE Signal Processing Lett.*, vol. 5, pp. 111–114, May 1998. DOI: 10.1109/97.668945 33

[29] G. H. Golub and C. F. Van Loan, *Matrix Computations*. Baltimore, MD: The Johns Hopkins University Press, 1996. 7, 8

[30] Y. Gu, J. Jin, and S. Mei, "ℓ_0 norm constraint LMS algorithm for sparse system identification," *IEEE Signal Processing Lett.*, vol. 16, pp. 774–777, Sept. 2009. DOI: 10.1109/LSP.2009.2024736 44, 45

[31] R. W. Harris, D. M. Chabries, and F. A. Bishop, "A variable step (VS) adaptive filter algorithm," *IEEE Trans. Acoust., Speech, Signal Processing*, vol. ASSP-34, pp. 309–316, Apr. 1986. DOI: 10.1109/TASSP.1986.1164814 31

[32] E. Hänsler and G. Schmidt, *Acoustic Echo and Noise Control–A Practical Approach*. Hoboken, NJ: Wiley, 2004. DOI: 10.1002/0471678406 16

[33] S. Haykin, *Adaptive Filter Theory*. Fourth Edition, Upper Saddle River, NJ: Prentice-Hall, 2002. 1, 2, 4, 15, 19, 25, 28, 29, 55, 56, 65, 67, 81

[34] S. I. Hill and R. C. Williamson, "Convergence of exponentiated gradient algorithms," *IEEE Trans. Signal Processing*, vol. 49, pp. 1208–1215, June 2001. DOI: 10.1109/78.923303 47, 50

[35] J. Homer, I. Mareels, R. R. Bitmead, B. Wahlberg, and A. Gustafsson, "LMS estimation via structural detection," *IEEE Trans. Signal Processing*, vol. 46, pp. 2651–2663, Oct. 1998. DOI: 10.1109/78.720368 4, 39, 55

[36] O. Hoshuyama, R. A. Goubran, and A. Sugiyama, "A generalized proportionate variable step-size algorithm for fast changing acoustic environments," in *Proc. IEEE ICASSP*, 2004, pp. IV-161–IV-164. DOI: 10.1109/ICASSP.2004.1326788 75, 96

[37] P. O. Hoyer, "Non-negative matrix factorization with sparseness constraints," *J. Machine Learning Res.*, vol. 49, pp. 1208–1215, June 2001. 10

[38] Y. Huang, J. Benesty, and J. Chen, *Acoustic MIMO Signal Processing*. Berlin, Germany: Springer-Verlag, 2006. 7, 10, 37

[39] P. J. Huber, *Robust Statistics*. New York: Wiley, 1981. DOI: 10.1002/0471725250 3

[40] B. Jeffs and M. Gunsay, "Restoration of blurred star field images by maximally sparse optimization," *IEEE Trans. Image Processing*, vol. 2, pp. 202–211, Apr. 1993. DOI: 10.1109/83.217223 45

[41] J. Kivinen and M. K. Warmuth, "Exponentiated gradient versus gradient descent for linear predictors," *Inform. Comput.*, vol. 132, pp. 1–64, Jan. 1997. DOI: 10.1006/inco.1996.2612 4, 37, 47, 48, 50, 90

[42] R. H. Kwong and E. W. Johnston, "A variable step size LMS algorithm," *IEEE Trans. Signal Processing*, vol. 40, pp. 1633–1642, July 1992. DOI: 10.1109/78.143435 31

[43] L. Liu, M. Fukumoto, and S. Zhang, "A variable parameter improved proportionate normalized LMS algorithm," in *Proc. IEEE APCCAS*, 2008, pp. 201–204. DOI: 10.1109/APCCAS.2008.4745995 42, 43, 44

[44] L. Liu, M. Fukumoto, S. Saiki, and S. Zhang, "A variable step-size proportionate affine projection algorithm for identification of sparse impulse response," *EURASIP J. Advances Signal Processing*, vol. 2009, article ID 150914, 10 pages, 2009. DOI: 10.1155/2009/150914 75

[45] P. Loganathan, A. W. H. Khong, and P. A. Naylor, "A class of sparseness-controlled algorithms for echo cancellation," *IEEE Trans. Audio, Speech, Language Processing*, vol. 17, pp. 1591–1601, Nov. 2009. DOI: 10.1109/TASL.2009.2025903 44, 59, 60, 61, 93

[46] X. Lu and B. Champagne, "Acoustic echo cancellation over a non-linear channel," in *Proc. IWAENC*, 2001, pp. 139–142. 3

[47] A. Mader, H. Puder, and G. U. Schmidt, "Step-size control for acoustic echo cancellation filters – An overview," *Signal Processing*, vol. 80, pp. 1697–1719, Sept. 2000. DOI: 10.1016/S0165-1684(00)00082-7 3, 31

[48] S. Makino, Y. Kaneda, and N. Koizumi, "Exponentially weighted step-size NLMS adaptive filter based on the statistics of a room impulse response," *IEEE Trans. Speech, Audio Processing*, vol. 1, pp. 101–108, Jan. 1993. DOI: 10.1109/89.221372 4, 55

[49] R. K. Martin, W. A. Sethares, R. C. Williamson, and C. R. Johnson, Jr., "Exploiting sparsity in adaptive filters," *IEEE Trans. Signal Processing*, vol. 50, pp. 1883–1894, Aug. 2002. DOI: 10.1109/TSP.2002.800414 54

[50] V. J. Mathews and Z. Xie, "A stochastic gradient adaptive filter with gradient adaptive step size," *IEEE Trans. Signal Processing*, vol. 41, pp. 2075–2087, June 1993. DOI: 10.1109/78.218137 31

[51] D. R. Morgan and S. G. Kratzer, "On a class of computationally efficient, rapidly converging, generalized NLMS algorithms," *IEEE Signal Processing Lett.*, vol. 3, pp. 245–247, Aug. 1996. DOI: 10.1109/97.511808 31

[52] Y. Murakami, M. Yamagishi, M. Yukawa, and I. Yamada, "A sparse adaptive filtering using time-varying soft-thresholding techniques," in *Proc. IEEE ICASSP*, 2010, pp. 3734–3737. 44

[53] T. Okuno, M. Fukushima, and M. Tohyama, "Adaptive cross-spectral technique for acoustic echo cancellation," *IEICE Trans. Fundamentals*, vol. E82-A, pp, 634–639, Apr. 1999. 3

[54] K. Ozeki and T. Umeda, "An adaptive filtering algorithm using an orthogonal projection to an affine subspace and its properties," *Electron. Commun. Jpn.*, vol. 67-A, pp. 19–27, May 1984. 2, 73, 81

[55] C. Paleologu, J. Benesty, and S. Ciochină, "A variable step-size affine projection algorithm designed for acoustic echo cancellation," *IEEE Trans. Audio, Speech, Language Processing*, vol. 16, pp. 1466–1478, Nov. 2008. DOI: 10.1109/TASL.2008.2002980 3, 80, 82, 100

[56] C. Paleologu, S. Ciochină, and J. Benesty, "Variable step-size NLMS algorithm for under-modeling acoustic echo cancellation," *IEEE Signal Processing Lett.*, vol. 15, pp. 5–8, 2008. DOI: 10.1109/LSP.2007.910276 3, 95

[57] C. Paleologu, J. Benesty, and S. Ciochină, "An improved proportionate NLMS algorithm based on the ℓ_0 norm," in *Proc. IEEE ICASSP*, 2010, pp. 309–312. 44, 45, 91

[58] C. Paleologu, S. Ciochină, and J. Benesty, "An efficient proportionate affine projection algorithm for echo cancellation," *IEEE Signal Processing Lett.*, vol. 17, pp. 165–168, Feb. 2010. DOI: 10.1109/LSP.2009.2035665 77, 99

[59] D. I. Pazaitis and A. G. Constantinides, "A novel kurtosis driven variable step-size adaptive algorithm," *IEEE Trans. Signal Processing*, vol. 47, pp. 864–872, Mar. 1999. DOI: 10.1109/78.747793 31

[60] B. D. Rao and K. Kreutz-Delgado, "An affine scaling methodology for best basis selection," *IEEE Trans. Signal Processing*, vol. 47, pp. 187–200, Jan. 1999. DOI: 10.1109/78.738251 45

[61] H. Rey, L. Rey Vega, S. Tressens, and J. Benesty, "Variable explicit regularization in affine projection algorithm: robustness issues and optimal choice," *IEEE Trans. Signal Processing*, vol. 55, pp. 2096–2108, May 2007. DOI: 10.1109/TSP.2007.893197 3

[62] L. Rey Vega, H. Rey, J. Benesty, and S. Tressens, "A family of robust algorithms exploiting sparsity in adaptive filters," *IEEE Trans. Audio, Speech, Language Processing*, vol. 17, pp. 572–581, May 2009. DOI: 10.1109/TASL.2008.2010156 3

[63] L. Rey Vega, H. Rey, and J. Benesty, "A robust variable step-size affine projection algorithm," *Signal Processing*, vol. 90, pp. 2806–2810, Sept. 2010. DOI: 10.1016/j.sigpro.2010.03.029 3

[64] S. G. Sankaran and A. A. L. Beex, "Convergence behavior of affine projection algorithms," *IEEE Trans. Signal Processing*, vol. 48, pp. 1086–1096, Apr. 2000. DOI: 10.1109/78.827542 82

[65] H.-C. Shin, A. H. Sayed, and W.-J. Song, "Variable step-size NLMS and affine projection algorithms," *IEEE Signal Processing Lett.*, vol. 11, pp. 132–135, Feb. 2004. DOI: 10.1109/LSP.2003.821722 3, 31

[66] M. M. Sondhi, "An adaptive echo canceler," *Bell Syst. Tech. J.*, vol. XLVI-3, pp. 497–510, Mar. 1967 1

[67] M. M. Sondhi, D. R. Morgan, and J. L. Hall, "Stereophonic acoustic echo cancellation – an overview of the fundamental problem," *IEEE Signal Processing Lett.*, vol. 2, pp. 148–151, Aug. 1995. DOI: 10.1109/97.404129 16

[68] F. das Chagas de Souza, O. J. Tobias, R. Seara, and D. R. Morgan, "A PNLMS algorithm with individual activation factors," *IEEE Trans. Signal Processing*, vol. 58, pp. 2036–2047, Apr. 2010. DOI: 10.1109/TSP.2009.2038420 61, 62, 63, 93

[69] A. Sugiyama, H. Sato, A. Hirano, and S. Ikeda, "A fast convergence algorithm for adaptive FIR filters under computational constraint for adaptive tap-position control," *IEEE Trans. Circuits Syst. II*, vol. 43, pp. 629–636, Sept. 1996. DOI: 10.1109/82.536759 4, 39, 55

[70] M. Tanaka, Y. Kaneda, S. Makino, and J. Kojima, "A fast projection algorithm for adaptive filtering," *IEICE Trans. Fundamentals*, vol. E78-A, pp. 1355–1361, Oct. 1995. 2, 78

[71] K. Wagner and M. Doroslovački, "Gain allocation in proportionate-type NLMS algorithms for fast decay of output error at all times," in *Proc. IEEE ICASSP*, 2009, pp. 3117–3120. DOI: 10.1109/ICASSP.2009.4960284 58, 59, 92

[72] S. Werner, J. A. Apolinário Jr., and P. S. R. Diniz, "Set-membership proportionate affine projection algorithms," *EURASIP J. Audio, Speech, Music Processing*, vol. 2007, no. 1, pp. 1–10, 2007. DOI: 10.1155/2007/34242 74, 75

[73] B. Widrow and M. E. Hoff, Jr., "Adaptive switching circuits," *IRE WESCON Conv. Rec.*, Pt. 4, 1960, pp. 96–104. 28

[74] B. Widrow, "Adaptive filters," in *Aspects of Network and System Theory*, R. E. Kalman and N. DeClaris, Eds., New York: Holt, Rinehart and Winston, 1970. 25, 28

[75] B. Widrow and S. D. Stearns, *Adaptive Signal Processing*. Englewood Cliffs, NJ: Prentice Hall, 1985. 29

[76] N. Wiener, *Extrapolation, Interpolation, and Smoothing of Stationary Time Series*. New York: John Wiley & Sons, 1949. 19

[77] Y. V. Zakharov and T. C. Tozer, "Multiplication-free iterative algorithm for LS problem," *IEE Electronics Lett.*, vol. 40, no. 9, pp. 567–569, Apr. 2004. DOI: 10.1049/el:20040353 78

[78] Y. V. Zakharov, "Low complexity implementation of the affine projection algorithm," *IEEE Signal Processing Lett.*, vol. 15, pp. 557–560, 2008. DOI: 10.1109/LSP.2008.2001111 78

[79] *Digital Network Echo Cancellers*, ITU-T Rec. G.168, 2002. 87

Index

Authors' Biographies

CONSTANTIN PALEOLOGU

Constantin Paleologu was born in Romania in 1975. In 1998 he received the M.S. degree in telecommunications networks from the Faculty of Electronics, Telecommunications, and Information Technology, University Politehnica of Bucharest, Romania. He also received a masters degree in digital signal processing in 1999, and a Ph.D. degree (MAGNA CUM LAUDE) in adaptive signal processing in 2003, both from the same institution. As a Ph.D. student (from December 1999 to July 2003), he worked on adaptive filters and echo cancellation. Since October 1998 he has been with the Telecommunications Department, University Politehnica of Bucharest where he is currently an Associate Professor. His research interests include adaptive signal processing, speech enhancement, and multimedia communications. He received the "IN HOC SIGNO VINCES" award from the Romanian National Research Council in 2009. He was the co-chair of the 2008 International Conference on Computing in the Global Information Technology (ICCGI). He is a member of the Steering Committee of the 2010 International Conference on Digital Telecommunications (ICDT). He serves as the Editor-in-Chief of the International Journal on Advances in Systems and Measurements. He is a Fellow of the International Academy, Research, and Industry Association (IARIA) since 2008.

JACOB BENESTY

Jacob Benesty was born in 1963. He received the Masters degree in microwaves from Pierre & Marie Curie University, France, in 1987, and the Ph.D. degree in control and signal processing from Orsay University, France, in April 1991. During his Ph.D. program (from November 1989 to April 1991), he worked on adaptive filters and fast algorithms at the Centre National d'Etudes des Telecommunications (CNET), Paris, France. From January 1994 to July 1995, he worked at Telecom Paris University on multichannel adaptive filters and acoustic echo cancellation. From October 1995 to May 2003, he was first a Consultant and then a Member of the Technical Staff at Bell Laboratories, Murray Hill, NJ, USA. In May 2003, he joined INRS-EMT, University of Quebec, in Montreal, Quebec, Canada, as a Professor. His research interests are in signal processing, acoustic signal processing, and multimedia communications. Dr. Benesty received the 2001 and 2008 Best Paper Awards from the IEEE Signal Processing Society. He was a member of the editorial board of the EURASIP Journal on Applied Signal Processing, a member of the IEEE Audio & Electroacoustics Technical Committee, the co-chair of the 1999 International Workshop on Acoustic Echo and Noise Control (IWAENC), and the general co-chair of the 2009 IEEE Workshop on

Applications of Signal Processing to Audio and Acoustics (WASPAA). Dr. Benesty co-authored and co-edited many books in the area of acoustic signal processing.

SILVIU CIOCHINĂ

Silviu Ciochină received the M.S. degree in electronics and telecommunications in 1971 and the Ph.D. degree in communications in 1978, both from the University Politehnica of Bucharest, Romania. From 1971 to 1979, he was an Assistant Professor, and from 1979 to 1995 a Lecturer at the University Politehnica of Bucharest, Faculty of Electronics, Telecommunications, and Information Technology. Since 1995, he has been a Professor at the same faculty. Since 2004, he has been the Head of the Telecommunications Department. His main research interests are in the areas of signal processing and wireless communications, including adaptive algorithms, spectrum estimation, fast algorithms, channel estimation, multi-antenna systems, and broadband wireless technologies. He received the "Traian Vuia" award in 1981 and the "Gheorghe Cartianu" award in 1997, both from the Romanian Academy.